METHODS IN MOLECULAR BIOLOGY

Series Editor
John M. Walker
School of Life Sciences
University of Hertfordshire
Hatfield, Hertfordshire, AL10 9AB, UK

For further volumes:
http://www.springer.com/series/7651

Cancer Genomics and Proteomics

Methods and Protocols

Second Edition

Edited by

Narendra Wajapeyee

Department of Pathology and Yale Cancer Center,
Yale University School of Medicine, New Haven, CT, USA

Editor
Narendra Wajapeyee
Department of Pathology and Yale Cancer Center
Yale University School of Medicine
New Haven, CT, USA

ISSN 1064-3745 ISSN 1940-6029 (electronic)
ISBN 978-1-4939-0991-9 ISBN 978-1-4939-0992-6 (eBook)
DOI 10.1007/978-1-4939-0992-6
Springer New York Heidelberg Dordrecht London

Library of Congress Control Number: 2014942924

Printed on acid-free paper

Humana Press is a brand of Springer
Springer is part of Springer Science+Business Media (www.springer.com)

Preface

Normal cells undergo multiple genetic and epigenetic alterations to become cancerous. These changes affect genome, transcriptome, and proteome of cancer cells. Therefore, to fully elucidate the mechanism of tumor initiation and progression as well as to identify targets for effective cancer therapy a comprehensive overview of tumor cells is necessary. In the last 10 years, cancer researchers have seen exceptional progress in high-throughput experimental approaches that makes analysis of cancer genome, transcriptome, and proteome within the reach of a standard research laboratory.

This edition of methods in molecular biology showcases some of the representative methods for the analyses of cancer genome and proteome that have illuminated us about the changes in cancer cells. Collectively, I anticipate that these protocols along with the other techniques described in this edition will allow readers to analyze various aspects of tumor initiation and progression.

New Haven, CT *Narendra Wajapeyee*

Contents

Contributors

KAVITHA K. ANANDALINGAM • *Department of Biomedical Engineering, Yale University, New Haven, CT, USA*
JAGADISH BELOOR • *Department of Internal Medicine, Section of Infectious Diseases, Yale University School of Medicine, New Haven, CT, USA*
FEDERICO BERNAL • *Metabolism Branch, Center for Cancer Research, National Cancer Institute, US National Institutes of Health, Bethesda, MD, USA*
STEPHANIE BLEICKEN • *Membrane Biophysics, Max Planck Institute for Intelligent Systems, Germany; Membrane Biophysics, German Cancer Research Center, Heidelberg, Germany; Membrane Biophysics, Interfaculty Institute of Biochemistry, Eberhard Karls University Tübingen, Tübingen, Germany*
CRAIG J. CEOL • *Program in Molecular Medicine, Department of Cancer Biology, Program in Cell and Developmental Dynamics, University of Massachusetts Medical School, Worcester, MA, USA*
RAMANA V. DAVULURI • *Center for Systems and Computational Biology, Molecular and Cellular Oncogenesis Program, The Wistar Institute, Philadelphia, PA, USA*
ANWESHA DEY • *Department of Molecular Biology, Genentech Inc., South San Francisco, CA, USA; Department of Protein Chemistry, Genentech Inc., South San Francisco, CA, USA*
PERRY EVANS • *Department of Pathology, Yale University School of Medicine, New Haven, CT, USA*
CHRISTINA M. FERRER • *Department of Biochemistry and Molecular Biology, Drexel University College of Medicine, Philadelphia, PA, USA*
SHAOJIAN GAO • *Medical Oncology Branch, Center for Cancer Research, National Cancer Institute, Bethesda, MD, USA*
ANA J. GARCÍA-SÁEZ • *Membrane Biophysics, Max Planck Institute for Intelligent Systems, Germany; Membrane Biophysics, German Cancer Research Center, Heidelberg, Germany; Membrane Biophysics, Interfaculty Institute of Biochemistry, Eberhard Karls University Tübingen, Tübingen, Germany*
PETER M. GLAZER • *Department of Genetics, Yale University School of Medicine, New Haven, CT, USA; Department of Therapeutic Radiology, Yale University School of Medicine, New Haven, CT, USA*
MICHAEL R. GREEN • *Program in Gene Function and Expression, University of Massachusetts Medical School, Worcester, MA, USA*
SUJUAN GUO • *Virginia Tech Carilion Research Institute, Roanoke, VA, USA*
YANWEN GUO • *Department of Genetics, Yale Stem Cell Center and Yale Cancer Center, Yale University, New Haven, CT, USA*
RAVI GUPTA • *Center for Systems and Computational Biology, Molecular and Cellular Oncogenesis Program, The Wistar Institute, Philadelphia, PA, USA*
MICHAEL E. HURWITZ • *Yale Cancer Center and Department of Medicine, Yale University School of Medicine, New Haven, CT, USA*

RYAN JENSEN • *Department of Therapeutic Radiology, Yale University School of Medicine, New Haven, CT, USA*
SAMUEL G. KATZ • *Department of Pathology, Yale University School of Medicine, New Haven, CT, USA*
JUDE KENDALL • *Cold Spring Harbor Laboratory, Cold Spring Harbor, NY, USA*
MEGAN C. KING • *Department of Cell Biology, Yale University School of Medicine, New Haven, CT, USA*
DONALD S. KIRKPATRICK • *Department of Protein Chemistry, Genentech Inc., South San Francisco, CA, USA*
YONG KONG • *Department of Molecular Biophysics and Biochemistry & W.M. Keck Foundation Biotechnology Resource Laboratory, Yale University School of Medicine, New Haven, CT, USA*
ALEXANDER KRASNITZ • *Cold Spring Harbor Laboratory, Cold Spring Harbor, NY, USA*
MICHAEL KRAUTHAMMER • *Department of Pathology, Yale University School of Medicine, New Haven, CT, USA*
PRITI KUMAR • *Department of Internal Medicine, Section of Infectious Diseases, Yale University School of Medicine, New Haven, CT, USA*
LIAM CHANGWOO LEE • *Medical Oncology Branch, Center for Cancer Research, National Cancer Institute, Bethesda, MD, USA*
SANG-KYUNG LEE • *Department of Bioengineering, Hanyang University, Seoul, South Korea*
BRYAN A. LELAND • *Department of Cell Biology, Yale University School of Medicine, New Haven, CT, USA*
QIUNING LI • *School of Life Science and Technology, Shanghai Tech University, Shanghai, China*
YIFEI LIU • *Yale Stem Cell Center and Department of Genetics, Yale School of Medicine, New Haven, CT, USA*
JUN LU • *Department of Genetics, Yale Stem Cell Center and Yale Cancer Center, Yale University, New Haven, CT, USA*
JI LUO • *Medical Oncology Branch, Center for Cancer Research, National Cancer Institute, Bethesda, MD, USA*
STEPHEN MASTRIANO • *Department of Genetics, Yale Stem Cell Center and Yale Cancer Center, Yale University, New Haven, CT, USA*
NICOLE A. MCNEER • *Department of Biomedical Engineering, Yale University, New Haven, CT, USA*
SUSAN F. MURPHY • *Virginia Tech Carilion Research Institute, Roanoke, VA, USA*
ARVINDHAN NAGARAJAN • *Department of Pathology, Yale University School of Medicine, New Haven, CT, USA*
HYE YEONG NAM • *Samyang Biopharmaceuticals Corporation, Daejeon, South Korea*
CORRIE A. PAINTER • *Program in Molecular Medicine, Department of Cancer Biology, Program in Cell and Developmental Dynamics, University of Massachusetts Medical School, Worcester, MA, USA*
SHARMISTHA PAL • *Center for Systems and Computational Biology, Molecular and Cellular Oncogenesis Program, The Wistar Institute, Philadelphia, PA, USA*
MAURICIO J. REGINATO • *Department of Biochemistry and Molecular Biology, Drexel University College of Medicine, Philadelphia, PA, USA*
ADELE S. RICCIARDI • *Department of Biomedical Engineering, Yale University, New Haven, CT, USA; Department of Genetics, Yale University School of Medicine, New Haven, CT, USA*

CHRISTINE RODEN • *Department of Pathology, Yale University School of Medicine, New Haven, CT, USA*
W. MARK SALTZMAN • *Department of Biomedical Engineering, Yale University, New Haven, CT, USA*
DONALD SHARON • *Department of Genetics, Stanford University School of Medicine, Stanford, CA, USA*
ZHI SHENG • *Virginia Tech Carilion Research Institute, Roanoke, VA, USA; Department of Biological Sciences and Pathobiology, Virginia-Maryland College of Veterinary Medicine, Blacksburg, VA, USA*
MICHAEL SNYDER • *Department of Genetics, Stanford University School of Medicine, Stanford, CA, USA*
ZITO TSENG • *Yale Stem Cell Center and Department of Genetics, Yale School of Medicine, New Haven, CT, USA*
NARENDRA WAJAPEYEE • *Department of Pathology and Yale Cancer Center, Yale University School of Medicine, New Haven, CT, USA*
TAO WU • *Yale Stem Cell Center and Department of Genetics, Yale School of Medicine, New Haven, CT, USA*
JIANSHENG WU • *Department of Protein Chemistry, Genentech Inc, South San Francisco, CA, USA*
ANDREW XIAO • *Yale Stem Cell Center and Department of Genetics, Yale School of Medicine, New Haven, CT, USA*
SHIHUA ZHANG • *National Center for Mathematics and Interdisciplinary Sciences, Academy of Mathematics and Systems Science, Chinese Academy of Sciences, Beijing, China*
MEI ZHONG • *Genomics Core facilities, Yale Stem Cell Center, Yale School of Medicine, New Haven, CT, USA*
XIANGHONG JASMINE ZHOU • *Program in Molecular and Computational Biology, University of Southern California, Los Angeles, CA, USA*

Chapter 1

Genome-Wide Mapping of RNA Pol-II Promoter Usage in Mouse Tissues by ChIP-Seq

Sharmistha Pal, Ravi Gupta, and Ramana V. Davuluri

Abstract

Chromatin immunoprecipitation (ChIP), using antibody against RNA Pol-II, followed by massive parallel sequencing (ChIP-seq) are invaluable techniques for genome-wide identification of alternative promoters and their patterns of use in different tissues, cell types, and/or developmental stages. However, the identification of promoters cannot be performed solely based on the presence of Pol-II enrichment on a genomic location because of its enrichment throughout the transcribed genomic region and lack of highly specific antibodies that can distinguish promoter-bound Pol-II from elongating Pol-II. In order to overcome this limitation, we developed a combined Pol-II ChIP-seq and bioinformatics promoter prediction approach to identify promoter regions and their activity in different mouse tissues. Here, we describe the integrative approach to identify alternative promoters in the mouse genome.

Key words Chromatin immunoprecipitation (ChIP), ChIP-seq, Promoter prediction, Alternative promoters

1 Introduction

The analyses of mammalian genomes by NextGen sequencing suggest that the majority of mammalian genes generate multiple transcripts and protein isoforms with distinct functional roles. This transcript diversity is generated, in part, through the use of alternative promoters [1] and alternative splicing [2], which produce pre-mRNA and mRNA isoforms (transcript/splice variants), respectively. The use of alternative promoters plays a fundamental role in regulating different isoforms of a gene, e.g., *LEF1*, *TP73*, *RUNX1*, and *MYC*, in various mammalian tissues and at different developmental stages. Moreover, the altered expression of transcript variants and protein isoforms for numerous genes is linked with disease and its prognosis, and cancer cells manipulate regulatory mechanisms to express specific isoforms that confer drug resistance and survival advantages [3].

Narendra Wajapeyee (ed.), *Cancer Genomics and Proteomics: Methods and Protocols*, Methods in Molecular Biology, vol. 1176, DOI 10.1007/978-1-4939-0992-6_1, © Springer Science+Business Media New York 2014

Therefore, it is important to evaluate the activity of primary and alternative gene promoters in various normal tissues and disease conditions. The advent of NextGen sequencing technologies and computational methods to support this technology has significantly improved our ability to annotate mammalian gene regulatory regions. Chromatin immunoprecipitation (ChIP) coupled with sequencing (ChIP-seq) [4] is enabling the genome-wide identification of alternative promoters and their patterns of use. However, the ChIP-seq approach needs to be applied with caution because of the inherent problems. For example, promoters cannot be identified solely by the presence of Pol-II enrichment on a genomic location because of its enrichment throughout the transcribed genomic region and lack of highly specific antibodies that can distinguish promoter-bound Pol-II from elongating Pol-II. In order to overcome this limitation we developed an integrative Pol-II ChIP-seq and bioinformatics promoter prediction approach to identify promoter regions and their activity in different mouse tissues. Here, we describe the approach to identify active promoters in mouse tissues.

The ChIP-seq protocol described here was modified from Lee et al. [5], which was developed for ChIP-chip (ChIP followed by microarray analysis). The modified protocol was successfully applied in our published studies to identify alternative promoters in different mouse tissues [6] and brain developmental stages [7].

2 Materials

2.1 ChIP-Seq

1. Surgical blade, homogenizer, centrifuge, magnetic bead stand, protein G- or A-coupled dynabeads, PCR purification kit (e.g., Qiagen), water bath, real-time PCR machine.
2. User must have access to NextGen sequencer (e.g., Illumina GA or HiSeq) either in-house or through core facility.
3. 1× PBS.
4. Cross-linking solution: 50 mM HEPES–KOH pH 7.5, 100 mM NaCl, 1 mM EDTA, 0.5 mM EGTA, and 11 % formaldehyde. Prepare solution before use.
5. 2.5 M Glycine solution pH 3.5.
6. Protease inhibitor stock solutions: 100 mM PMSF, 2 mg/ml pepstatin A, 5 mg/ml leupeptin, 6 U/ml aprotinin.
7. Cell lysis buffer I: 50 mM HEPES–KOH pH 7.5, 140 mM NaCl, 1 mM EDTA, 10 % glycerol (v/v), 0.5 % NP-40 (v/v), 0.25 % Triton X-100 (v/v), 0.5 mM PMSF, 25 μg/ml pepstatin A, 25 μg/ml leupeptin, and 0.6 U/ml aprotinin. Add protease inhibitors before use. Store buffer at 4 °C.

8. Cell lysis buffer II: 10 mM Tris–HCl pH 8.0, 200 mM NaCl, 1 mM EDTA, 0.5 mM EGTA, 0.5 mM PMSF, 25 μg/ml pepstatin A, 25 μg/ml leupeptin, and 0.6 U/ml aprotinin. Add protease inhibitors before use. Store buffer at 4 °C.
9. Cell lysis buffer III: 10 mM Tris–HCl pH 8.0, 100 mM NaCl, 1 mM EDTA, 0.5 mM EGTA, 0.5 % N-lauroylsarcosine (w/v), 0.1 % sodium deoxycholate (w/v), 0.5 mM PMSF, 25 μg/ml pepstatin A, 25 μg/ml leupeptin, and 0.6 U/ml aprotinin. Add protease inhibitors before use. Store buffer at 4 °C.
10. Blocking buffer: 1× PBS, 0.5 % BSA. Prepare solution before use and keep on ice.
11. ChIP wash buffer I: 50 mM HEPES–KOH pH 7.5, 500 mM LiCl, 1 mM EDTA, 1.0 % NP-40 (v/v), 0.7 % sodium deoxycholate (w/v). Store buffer at 4 °C.
12. ChIP wash buffer II: 10 mM Tris–HCl pH 8.0, 50 mM NaCl, 1 mM EDTA. Store buffer at 4 °C.
13. ChIP elution buffer: 50 mM Tris–HCl pH 8.0, 10 mM EDTA, 1% SDS (w/v). Store buffer at room temperature.
14. RNAse solution: 10 mg/ml.
15. Proteinase K solution: 20 mg/ml.
16. Primers (user defined) and mastermix for real-time PCR.

2.2 ChIP-Seq Data Analysis

User must have access to a computer with Internet access, e.g., a PC running Microsoft Windows or Linux, an Apple Macintosh, or a Linux/UNIX workstation. The user should be familiar with the use of Internet browser (e.g., Mozilla Firefox, Google Chrome, Microsoft Internet Explorer), Python programming language, and R statistical package. If R programming package is not readily available the user can download the R base package from R-project website (via http://CRAN.R-project.org). The classification packages "rpart" and "randomForest" should be downloaded and installed in R. The list of commonly used efficient programs that aligns relatively short nucleotide sequences against a long reference sequence, such as the human genome, and ChIP-seq peak finding prediction programs are provided in Table 1. The R package implementation of supervised learning algorithm to discriminate promoter-associated Pol-II enrichment from enrichment elsewhere in the genome in ChIP-seq profiles should be downloaded from http://mpromdb.wistar.upenn.edu. Further, the user should be familiar with the use of UCSC genome browser (http://genome.ucsc.edu/) for uploading and visualizing the data.

Table 1
Web URLs of ChIP-seq data analysis programs

Program name (reference)	Description	Web URL
MPromDb [15]	Database of mammalian gene promoters	http://mpromdb.wistar.upenn.edu/
Bowtie [9]	Ultrafast, memory-efficient short read aligner	http://bowtie-bio.sourceforge.net/index.shtml
BWA [16]	Burrows–Wheeler Aligner (BWA) is an efficient program that aligns relatively short nucleotide sequences against a long reference sequence	http://bio-bwa.sourceforge.net/
SHRiMP2 [17]	Software package for aligning short reads against a target genome	http://compbio.cs.toronto.edu/shrimp/
Stampy [18]	Package for mapping of short reads onto a reference genome	http://www.well.ox.ac.uk/project-stampy
Promoter Prediction Program [13]	Algorithm to discriminate promoter-associated Pol-II enrichment from enrichment elsewhere in the genome in ChIP-seq profiles	http://mpromdb.wistar.upenn.edu/
MOSAiCS [10]	Flexible mixture model for detecting peaks in both one- and two-sample analyses of ChIP-seq data	http://www.stat.wisc.edu/~keles/Software/mosaics/index.html
MACS [11]	A ChIP-seq peak-finding algorithm that uses a dynamic Poisson distribution to effectively capture local biases in the genome sequence	http://liulab.dfci.harvard.edu/MACS
QuEST [12]	A kernel density estimator-based package for analysis of ChIP-seq data	http://mendel.stanford.edu/SidowLab/downloads/quest/
HOMER [14]	HOMER contains *annotatePeaks.pl* program for performing ChIP-seq peak annotation	http://biowhat.ucsd.edu/homer/ngs/annotation.html

3 Methods

3.1 Chromatin Immunoprecipitation and Massive Parallel Sequencing (ChIP-Seq)

This protocol described here was modified from Lee et al. [5], which was developed for ChIP-chip. This was successfully used in our published studies to identify alternative promoters in different mouse tissues [6] and brain developmental stages [7].

Add protease inhibitors before use, and prepare solutions fresh where indicated in Subheading 2.1. Keep samples cold at all times unless specified otherwise.

3.1.1 Chromatin Preparation

1. Mince finely about 1 g of freshly dissected mouse tissue (e.g., brain, kidney, liver, lung, or spleen) on ice and transfer to 50 ml tubes.
2. Add five times the volume of 1× PBS, and cross-link the minced tissue by adding 1/10th volume of cross-linking solution for 10 min at room temperature on a nutator.
3. To stop cross-linking add glycine to a final concentration of 0.125 M, mix by swirling, and place samples on ice.
4. Homogenize the samples on ice using a mechanical homogenizer till no tissue pieces are visible.
5. Spin samples at 1,200 ×*g* for 5 min at 4 °C. Discard supernatant, and wash cell pellet by suspending the pellet in 45 ml of 1× PBS. Collect cells by centrifugation at 1,200 ×*g* for 5 min at 4 °C.
6. Resuspend the cell pellet in 50 ml of cell lysis buffer I and place on nutator at 4 °C for 15 min. Check cell lysis by trypan blue staining on 5 μl of sample. If lysis is less than 80 % repeat the above step once more after collecting the pellet. Centrifuge the sample at 1,500 ×*g* for 5 min at 4 °C. Discard supernatant.
7. Resuspend the pellet in 50 ml of cell lysis buffer II and place on nutator at room temperature for 10 min. Centrifuge the sample at 1,500 ×*g* for 5 min at 4 °C. Discard supernatant.
8. Resuspend the pellet in 10 ml of cell lysis buffer III and place on nutator at 4 °C for 10 min. Sonicate sample to fragment cross-linked chromatin to a range of 0.2–0.5 kb. Sonication parameters like time, duty cycle, and output have to be optimized on the specific sonicator being used and for specific type of sample. For freshly dissected mouse tissues, acceptable sonication was achieved on a bioruptor with five cycles of 5 min (30 s ON and 1 min OFF). Keep samples cold at all times.
9. Add Triton X-100 to a final concentration of 1 %, mix, and centrifuge the sample at 17,900 ×*g* for 10 min at 4 °C.
10. Transfer the supernatant containing the solubilized chromatin to a 15 ml tube, and discard the pellet, which are cellular

debris. Keep aside 50 μl aliquot of the chromatin sample, freeze the rest in liquid nitrogen, and store at −80 °C.

11. Add 150 μl of elution buffer to the aliquot, and reverse cross-link at 65 °C for 6–14 h. Cool sample to room temperature, and add 200 μl TE pH 8.0 and 8 μl of RNAse solution. Incubate at 37 °C for 2 h. Add 4 μl of Proteinase K solution, mix, and incubate at 55 °C for another 2 h. Purify the DNA over PCR purification column as per the manufacturer's instructions.
12. Analyze the fragmentation of chromatin on a 1 % agarose gel. Majority of the DNA should be in 0.2–0.5 kb. Do not proceed with ChIP if desired fragmentation is not achieved.

3.1.2 Chromatin Immunoprecipitation (See ***Note 1****)*

13. Thaw the frozen chromatin on ice (2–3 h).
14. In the meantime, transfer 100 μl of protein G/A dynabeads to an Eppendorf tube and wash by inverting the tube 5–6 times with 1.5 ml blocking buffer. Prepare one tube per immunoprecipitation (control antibody, e.g., nonspecific IgG, anti-Pol-II antibody, and one extra tube to pre-clear 10 ml of chromatin). Collect beads on the magnetic rack, and discard supernatant. Repeat the above step twice more.
15. Resuspend the beads in 250 μl of blocking buffer, add 10 μg of antibody (ChIP grade), simultaneously block the bead surface, and bind the antibody on nutator for 4–6 h at 4 °C. Do not add any antibody to the beads that will be used for pre-clearing.
16. Collect the beads being blocked for pre-clearing, and transfer them to the thawed chromatin. Place the chromatin-containing tube on nutator for 2 h at 4 °C to pre-clear chromatin. Centrifuge the sample at 17,900 × *g*, and equally distribute the supernatant of pre-cleared chromatin to two 15 ml tubes.
17. Wash the antibody-bound beads twice with 1.5 ml of blocking solution and transfer to the pre-cleared chromatin tube. Incubate the samples on nutator at 4 °C for 14–16 h.
18. Collect the beads that have the bound antibody–chromatin complex by sequentially transferring 1.5 ml of sample and discarding the supernatant, which is the antibody-depleted chromatin.
19. Wash the bound nucleoprotein complexes extensively (inverting tubes 7–8 times) six times with 1 ml ChIP wash buffer I and once with 1 ml ChIP wash buffer II.
20. Elute the bound nucleoprotein complexes by adding 200 μl of elution buffer, vortexing the samples for 1 min, and incubating the tubes at 65 °C for 20 min. Centrifuge samples at 17,900 × *g* for 5 min at room temperature, transfer the eluted chromatin to a new tube, and incubate the sample to reverse cross-link at 65 °C for 6–14 h.

21. Proceed with RNAse, Proteinase K treatment, and DNA purification as before in **step 11**. Quantify the ChIP-enriched DNA using a picogreen assay (e.g., Quant-IT pico green dsDNA assay kit).

3.1.3 ChIP-Enriched DNA Quality Check and Sequencing

22. Check the quality of ChIP experiment by qPCR on known target genes (e.g., housekeeping gene promoters, body and 3′ end for anti-Pol-II) and nontarget regions (e.g., gene desert region for anti-Pol-II). 20–25 pg of ChIP-enriched DNA and input chromatin DNA (**step 11**) is analyzed by real-time PCR using specific primers, and fold enrichment is calculated relative to input DNA (Fig. 1). Proceed with sequencing of the ChIP-enriched DNA only if high enrichment is observed for positive targets and no enrichment is achieved on nontarget regions.
23. Process the purified DNA (5–10 ng, preferentially use same DNA amounts for the control and antigen-specific antibody) according to the instructions to prepare the library for sequencing ChIP-enriched DNA (e.g., Illumina Inc. ChIP-seq library prep kit). Sequencing of the library on the NexGen sequencer generates reads of length 36–75 bp (based on the user-specified sequencing parameters) for each sequenced molecule, and these reads are further processed as described below.

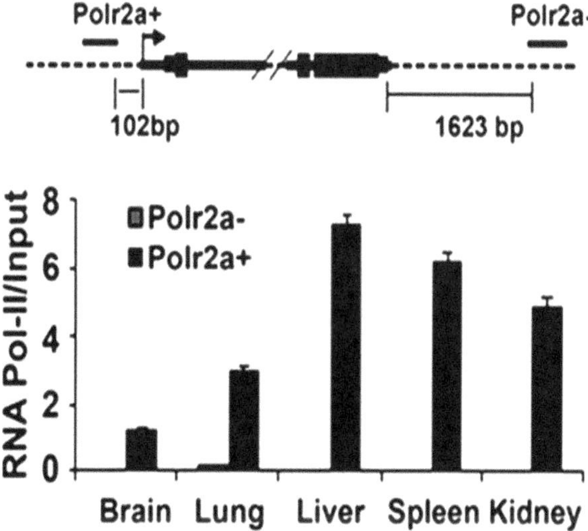

Fig. 1 Real-time PCR is used to evaluate the enrichment of Pol-II on the target region (Polr2a promoter-Polr2a+) and nontarget region (downstream of Polr2a gene-Polr2a–) on five mouse tissues. The enrichment of Pol-II is represented with respect to input chromatin. There is no enrichment observed on the nontarget region. (The figure has been taken from Sun et al. [6])

3.2 ChIP-Seq Data Processing

The approach described here is similar to the one used by previous published studies [4, 8] for ChIP-seq data analysis. Briefly, the analysis involves the following steps:

1. *Read mapping*: Use Bowtie [9] or any other program (Table 1) for alignment of sequence reads to the reference genome.
2. *Peak identification*: Identify statistically significant sequence read-enriched genomic regions (of length 1 kbp) by MOSAiCS [10], MACS [11], and QuEST [12] package (Table 1). A region can be considered statistically significant if the number of reads within that region is higher than the number expected due to random background.

3.2.1 Identification and Annotation of Pol-II Promoter Peaks

1. To identify the Pol-II-bound promoters from the ChIP-seq data, apply the promoter prediction program to discriminate Pol-II enrichment peak associated with promoter region from peaks associated with non-promoter region [13] (*see* **Note 2**).
2. For annotating the predicted promoters, refer to the gene information tracks from various sources available at UCSC genome browser. Apply the annotatePeaks.pl program in HOMER software for performing peak annotation [14].

4 Notes

1. DNA fragmentation, antibody, and quality of ChIP-enriched DNA are most important for the success of a ChIP sequencing experiment. Sonication is one of the most variable steps and has to be optimized for each sample and sonification equipment type. Antibodies used should be of ChIP grade. Finally, before proceeding with sequencing the enrichment on target regions and lack of enrichment on nontarget regions should be verified. It is advisable to test multiple regions before proceeding with sequencing library preparation.
2. Despite great progress, computational methods to effectively integrate ChIP-seq profiles to identify and annotate the promoter usage in specific cell/tissue types or developmental stages are still limited. The promoter prediction program suggested in this methodology was at best 95 % accurate in predicting a given ChIP-seq peak as promoter or non-promoter based on cross validation and testing on an independent data set. We suggest using a combination of different programs and databases, such as MPromDb [15], and take the consensus predictions for considering the reliable predictions. But, we expect some amount of noise in terms of false predictions and missed real promoters within a given genomic locus.

References

1. Davuluri RV, Suzuki Y, Sugano S, Plass C, Huang TH (2008) The functional consequences of alternative promoter use in mammalian genomes. Trends Genet 24:167–177
2. Wang ET, Sandberg R, Luo S, Khrebtukova I, Zhang L, Mayr C, Kingsmore SF, Schroth GP, Burge CB (2008) Alternative isoform regulation in human tissue transcriptomes. Nature 456:470–476
3. Pal S, Gupta R, Davuluri RV (2012) Alternative transcription and alternative splicing in cancer. Pharmacol Ther 136:283–294
4. Robertson G, Hirst M, Bainbridge M, Bilenky M, Zhao Y, Zeng T, Euskirchen G, Bernier B, Varhol R, Delaney A et al (2007) Genome-wide profiles of STAT1 DNA association using chromatin immunoprecipitation and massively parallel sequencing. Nat Methods 4:651–657
5. Lee TI, Johnstone SE, Young RA (2006) Chromatin immunoprecipitation and microarray-based analysis of protein location. Nat Protoc 1:729–748
6. Sun H, Wu J, Wickramasinghe P, Pal S, Gupta R, Bhattacharyya A, Agosto-Perez FJ, Showe LC, Huang TH, Davuluri RV (2011) Genome-wide mapping of RNA Pol-II promoter usage in mouse tissues by ChIP-seq. Nucleic Acids Res 39:190–201
7. Pal S, Gupta R, Kim H, Wickramasinghe P, Baubet V, Showe LC, Dahmane N, Davuluri RV (2011) Alternative transcription exceeds alternative splicing in generating the transcriptome diversity of cerebellar development. Genome Res 21:1260–1272
8. Zhang ZD, Rozowsky J, Snyder M, Chang J, Gerstein M (2008) Modeling ChIP sequencing in silico with applications. PLoS Comput Biol 4:e1000158
9. Shen R, Mo Q, Schultz N, Seshan VE, Olshen AB, Huse J, Ladanyi M, Sander C (2012) Integrative subtype discovery in glioblastoma using iCluster. PLoS One 7:e35236
10. Kuan PF, Chung DJ, Pan GJ, Thomson JA, Stewart R, Keles S (2011) A statistical framework for the analysis of ChIP-Seq data. J Am Stat Assoc 106:891–903
11. Zhang Y, Liu T, Meyer CA, Eeckhoute J, Johnson DS, Bernstein BE, Nussbaum C, Myers RM, Brown M, Li W et al (2008) Model-based analysis of ChIP-Seq (MACS). Genome Biol 9:R137
12. Valouev A, Johnson DS, Sundquist A, Medina C, Anton E, Batzoglou S, Myers RM, Sidow A (2008) Genome-wide analysis of transcription factor binding sites based on ChIP-Seq data. Nat Methods 5:829–834
13. Gupta R, Wikramasinghe P, Bhattacharyya A, Perez FA, Pal S, Davuluri RV (2010) Annotation of gene promoters by integrative data-mining of ChIP-seq Pol-II enrichment data. BMC Bioinformatics 11(Suppl 1):S65
14. Heinz S, Benner C, Spann N, Bertolino E, Lin YC, Laslo P, Cheng JX, Murre C, Singh H, Glass CK (2010) Simple combinations of lineage-determining transcription factors prime cis-regulatory elements required for macrophage and B cell identities. Mol Cell 38:576–589
15. Gupta R, Bhattacharyya A, Agosto-Perez FJ, Wickramasinghe P, Davuluri RV (2011) MPromDb update 2010: an integrated resource for annotation and visualization of mammalian gene promoters and ChIP-seq experimental data. Nucleic Acids Res 39:D92–D97
16. Li H, Durbin R (2010) Fast and accurate long-read alignment with Burrows-Wheeler transform. Bioinformatics 26:589–595
17. David M, Dzamba M, Lister D, Ilie L, Brudno M (2011) SHRiMP2: sensitive yet practical SHort Read Mapping. Bioinformatics 27:1011–1012
18. Lunter G, Goodson M (2011) Stampy: a statistical algorithm for sensitive and fast mapping of Illumina sequence reads. Genome Res 21:936–939

Chapter 2

Using Native Chromatin Immunoprecipitation to Interrogate Histone Variant Protein Deposition in Embryonic Stem Cells

Zito Tseng, Tao Wu, Yifei Liu, Mei Zhong, and Andrew Xiao

Abstract

Chromatin immunoprecipitation combined with massive parallel sequencing (ChIP-Seq) is a powerful epigenetics technique for interrogating the genome-wide localization of histone modifications, histone variants, and other chromatin-associating factors. In brief, chromatin pellets are fractionated from the nuclei, and then fragmented by enzymatic digestion or sonication. Chromatin regions associated with proteins of interest are enriched by immunoprecipitation with specific antibodies. After the immunoprecipitation, DNA fragments are extracted, amplified during sequencing library construction, and sequenced by high-throughput sequencing. Here, we describe the native chromatin immunoprecipitation of a rare histone variant, H2A.X, followed by high-throughput sequencing, in mouse embryonic stem cells.

Key words Native chromatin immunoprecipitation (native ChIP), Chromatin immunoprecipitation combined with massive parallel sequencing (ChIP-seq), Epigenetics, Histone variants, H2A.X, Embryonic stem cells

1 Introduction

Chromatin immunoprecipitation combined with massive parallel sequencing (ChIP-Seq) has become a critical method in the field of epigenetics for identifying the genomic regions associated with histone modifications, histone variants, and regulatory factors for histones and DNA. Since ChIP was first develop in 1984 to study the genomic regions associated with RNA polymerase [1], there have been many modifications to optimize the original protocol [2–5]. Regardless of the variations of different types of ChIP, these protocols can be broken down into four major steps: Fractionation and isolation of chromatin pellets from the nuclei, fragmentation of the chromatin pellets, purification and enrichment of genomic regions of interest from the fragmented chromatin pellets by immunoprecipitation, and isolation of DNA associated with the genomic regions of interest. The isolated ChIP DNA is further

Narendra Wajapeyee (ed.), *Cancer Genomics and Proteomics: Methods and Protocols*, Methods in Molecular Biology, vol. 1176, DOI 10.1007/978-1-4939-0992-6_2,

used in library construction and high-throughput sequencing in the ChIP-seq experiments (Fig. 1a).

1.1 N-ChIP and X-ChIP

Two major types of ChIP protocols have been developed, native ChIP (N-ChIP) and cross-linking ChIP (X-ChIP). N-ChIP retains the native chromatin configuration, while the X-ChIP uses formaldehyde to cross-link DNA with associated proteins [3, 5].

They each have advantages and disadvantages [6]. First, N-ChIP is not applicable to most non-histone proteins because without cross-linking, DNA cannot be co-immunoprecipitated with proteins that do not bind DNA tightly. Second, N-ChIP does not cross-link DNA and proteins, and thus, histones might slide on the genome during the sample preparation. This may be problematic if high resolution is of priority in one's research. For example, high resolution is critical for identifying DNA motifs associated with the protein of interest. However, this concern becomes trivial if the protein of interest occupies large chromatin domains (>100 kb) because sliding on DNA molecules usually occurs at small scales. Third, nuclease digestion performed in N-ChIP may be biased towards certain chromatin regions. For instance, open chromatin regions can be digested more easily than heterochromatin regions [7].

On the other hand, since antibodies are usually raised against unfixed epitopes, N-ChIP is usually much more efficient than

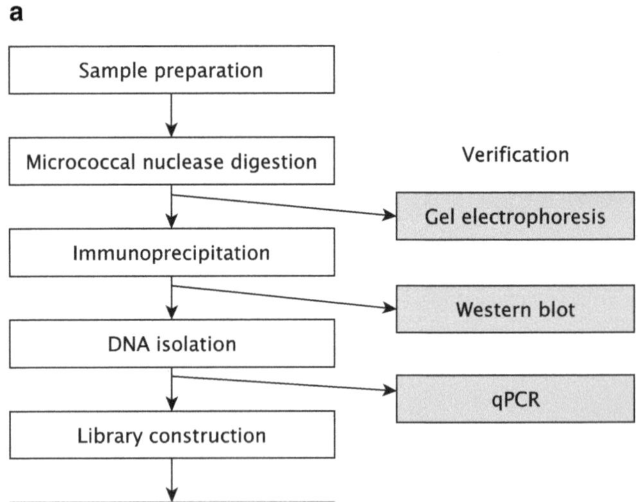

Fig. 1 Flow diagrams of the N-ChIP-Seq protocol. (**a**) Major steps of the N-ChIP include sample preparation, micrococcal nuclease (MNase) digestion, immunoprecipitation, DNA isolation, Library construction, and high-throughput sequencing. Among these steps, the MNase digestion can be verified by gel electrophoresis, and the immunoprecipitation can be verified by western blot and qRT-PCR. (**b**) A more detailed flow chart of the protocol described in this chapter, N-ChIP of H2A.X in ESC

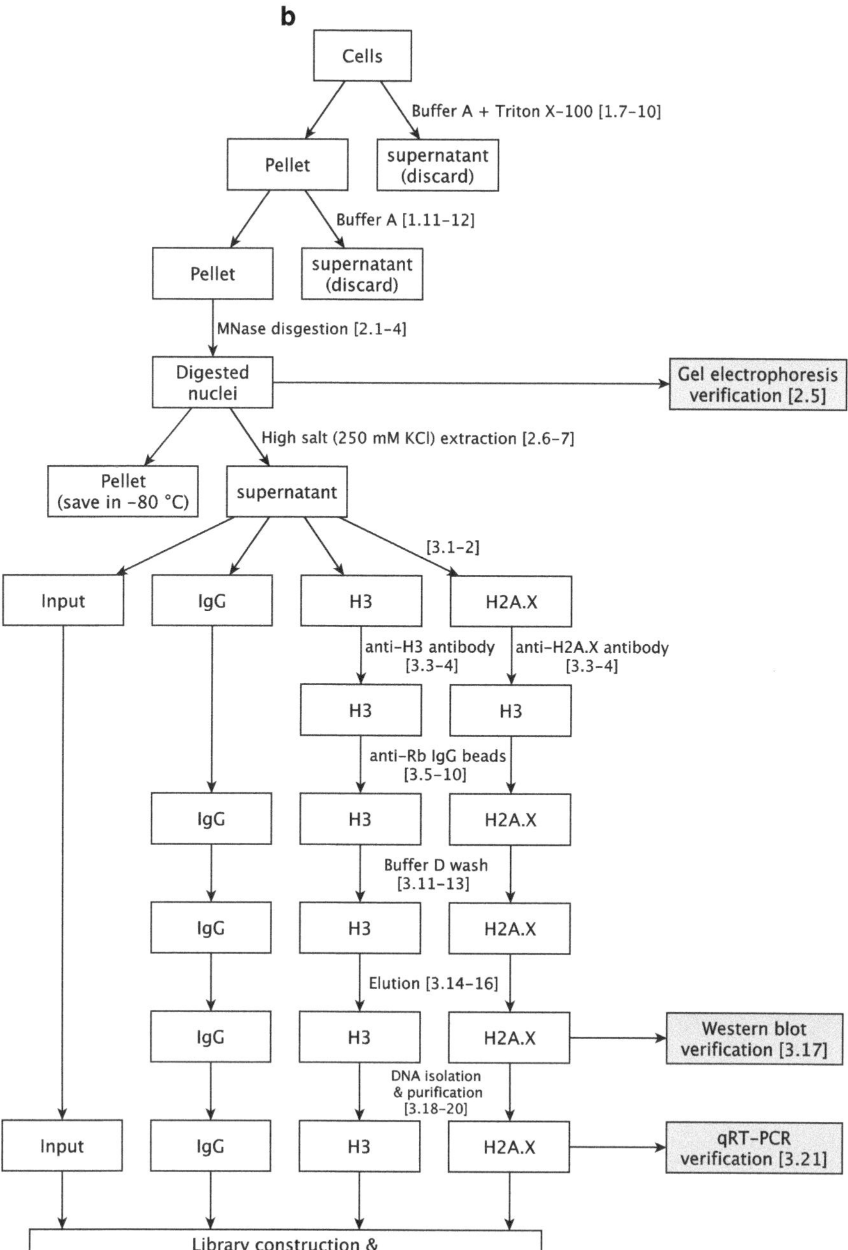

Fig. 1 (continued)

X-ChIP. In addition, cross-linking often stabilizes transient or nonspecific interactions, and therefore, increases the background noise. Based on these considerations, we favor the view that proteins with direct and strong interactions with DNA, such as histone modifications, histone variants, and chromatin remodeling factors, should be ideal targets for N-ChIP.

1.2 Controls and Verifications

The success of ChIP experiments largely depends on the quality of the antibodies and the abundance of protein of interest [8]. Thus, it is important to include negative controls, whenever it is possible, to filter out nonspecific signals. In addition, the qualities of the commercially available ChIP grade antibodies are highly variable [8]. Some commercial antibodies failed in ChIP experiments [9]. Therefore, it is important to verify the specificity of the antibody by examining the immunoprecipitated samples by western blotting and examining the isolated DNA by qPCR before a large-scale experiment.

To normalize the ChIP signals, it is important to include an input DNA control and a H3 ChIP control. First, the percentage of the ChIP signals over the DNA input signals shows the relative occupancy of the locus by the protein of interest in the ChIP experiments. This percentage usually varies from 0.1 to 1 %. Second, the percentage of the ChIP signals over the H3 signals shows the relative enrichment of the factor of interest. This is an important control as the H3 signals reflect the nucleosome occupancy. For example, an observed change for a histone modification may reflect a general fluctuation in nucleosome occupancy rather than specific alterations in the activities of the enzymes that modify this mark. Finally, nonspecific antibodies, such as IgG, can also be included to examine the specificity of the ChIP.

Here, we adapted a chromatin extraction approach [10] to perform N-ChIP of a rare histone variant, H2A.X, in mouse embryonic stem cells (ESC) (Fig. 1b). Using H2A.X KO ESC as control, we enriched and isolated the genomic DNA specifically associated with H2A.X nucleosomes. We also verified the specificity of the N-ChIP by western blot and qRT-PCR. The library construction following the N-ChIP was adapted from a protocol by Illumina Inc. [11]. The ensuing high-throughput sequencing provides an accurate description of the genome-wide localization of H2A.X in ESC (data not shown).

2 Material

2.1 Sample Preparation for N-ChIP

1. 1× PBS.
2. Buffer A:

   10 mM Hepes (pH 7.9) (Sigma-Aldrich, H3375).

   10 mM KCl (JT Baker, 3040).

   1.5 mM $MgCl_2$ (JT Baker, 2444).

   0.34 M Sucrose (Sigma-Aldrich, S0389).

   10 % Glycerol (American Bioanalytical, AB00751).

   Protease inhibitor (Roche, 13318000) (*see* **Note 1**).
3. 20 % Triton X-100 (Sigma-Aldrich, T8787).

2.2 Micrococcal Nuclease Digestion

1. 1 M $CaCl_2$.
2. Micrococcal nuclease.
3. 0.1 M EGTA.
4. Phenol–chloroform–isoamyl alcohol mixture.
5. 100 bp DNA Ladder.
6. Agarose.
7. 2 M KCl.

2.3 Immunoprecipitation

1. 20 % Triton X-100.
2. Anti-H2A.X antibody (Abcam).
3. Anti-H3 antibody (Abcam).
4. Dynabeads M-280 Sheep Anti-Rabbit IgG (Invitrogen).
5. DynaMag-2 Magnet (Invitrogen).
6. 0.5 % BSA in PBS.
7. Buffer D:
   20 mM HEPES (pH 7.9).
   0.2 mM EDTA.
   250 mM KCl.
   20 % Glycerol.
   0.2 % Triton X-100 (*see* **Note 2**).
8. Elution buffer:
   50 mM Tris–HCl (pH 8.0).
   10 mM EDTA.
   1 % SDS.
9. Thermomixer compact.
10. MinElute Reaction Cleanup Kit.
11. Nuclease-free water.

2.4 Library Construction

1. End-It DNA End-Repair Kit.
2. QIAquick PCR Purification Kit.
3. Nuclease-free water.
4. 1 mM dATP.
5. NEBuffer 2.
6. Klenow Fragment ($3' \rightarrow 5'$ exo-) (New England).
7. MinElute Reaction Cleanup Kit.
8. LigaFast Rapid DNA Ligation System (Promega).
9. TruSeq DNA LT Sample Prep Kit v2 (Illumina).
10. 10× BlueJuice Gel Loading Buffer (Invitrogen).
11. E-Gel EX Gel, 2 % (Invitrogen).

12. Clean scalpels.
13. QIAquick Gel Extraction Kit (QIAGEN).
14. Phusion High-Fidelity PCR Master Mix (New England Biolabs).
15. NanoDrop 2000 (Thermo scientific).
16. 2100 Bioanalyzer (Agilent).
17. HiSeq 2000 (Illumina).

3 Methods

3.1 Sample Preparation for N-ChIP

1. Culture ESC on 3 150 mm dishes to 60–80 % confluency (*see* **Notes 3** and **4**).
2. Remove the cell culture media.
3. Wash each dish with 15 mL ice-cold PBS twice.
4. Add 3 mL ice-cold PBS to each dish.
5. Use a cell scraper to scrap the dishes thoroughly. Gather the cell suspension from 3 dishes into a 15 mL tube.
6. Centrifuge for 5 min at 1,300×*g* at 4 °C. Discard the supernatant.
7. Resuspend the pellet in 1.5 mL buffer A (*see* **Note 5**). Pipet the sample to a 1.5 mL tube.
8. Add 3 μL 20 % Triton X-100 (final concentration: 0.1 %). Invert 4 times gently (*see* **Note 6**).
9. Lay on ice for 7 min (*see* **Note 7**).
10. Centrifuge for 5 min at 1,300×*g* at 4 °C. Discard the supernatant.
11. Resuspend the pellet in 1.5 mL buffer A with wide-orifice tips (*see* **Note 8**).
12. Centrifuge for 5 min at 1,300×*g* at 4 °C. Discard the supernatant.
13. Resuspend the pellet in 1 mL buffer A with wide-orifice tips (*see* **Note 9**).

3.2 Micrococcal Nuclease Digestion

1. Add 10 μL 1 M $CaCl_2$ (final concentration 1 mM) (*see* **Note 10**).
2. Prewarm the tube in 37 °C water bath for 5 min and get 0.1 M EGTA solution ready.
3. Add 10 μL micrococcal nuclease and incubate the tube in 37 °C water bath for 2 min. Vortex briefly (*see* **Note 11**).
4. Immediately put the tube on ice and add 10 μL 0.1 M EGTA (final concentration 100 μM). Mix thoroughly by vortexing.

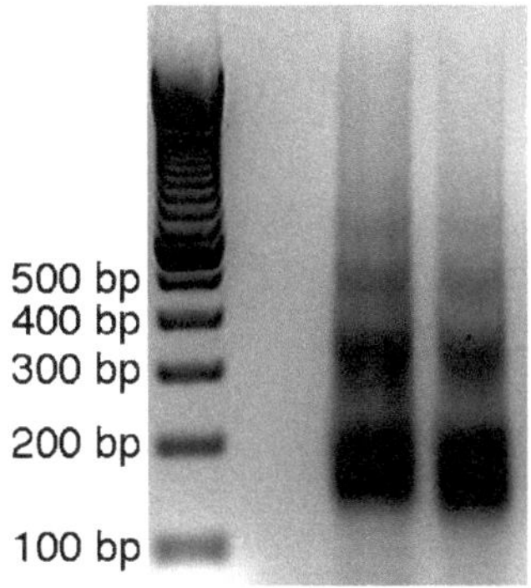

Fig. 2 After micrococcal nuclease digestion, the DNA fragments extracted from the mononucleosomes are about 150–200 base pairs long. Left: 100 bp DNA Ladder. *Middle and Right*: DNA extracted from micrococcal nuclease digested chromatins

3.2.1 Verify Micrococcal Nuclease Digestion

1. Aliquot 10 μL sample to a new 1.5 mL tube. Add 20 μL water and 30 μL phenol–chloroform.
2. Close the tube tightly and shake it vigorously for 15 s.
3. Centrifuge for 5 min at 20,000 × *g*.
4. Pipette 20 μL of the upper aqueous phase to a new tube. Perform DNA gel electrophoresis with 2 % agarose gel for 40 min at 100 V. Examine whether the size of the DNA fragments is mostly between 150 and 200 base pairs (Fig. 2) (*see* **Note 12**).
5. Add 150 μL 2 M KCl to the sample from **step 4** of Subheading 3.1.1 nuclear chromatin extraction (final concentration 250 mM). Rotate at 4 °C for 2 h.
6. Centrifuge for 5 min at 21,000 × *g* at 4 °C. Move the supernatant to a new tube.

3.3 Immunoprecipitation

1. Add 6 μL 20 % Triton X-100 (final concentration 0.1 %).
2. Save 30 μL sample from **step 6** of Subheading 3.2.1 as "input." Aliquot 200 μL sample to a new tube, "IgG." Aliquot 200 μL sample to a new tube, "H3." Aliquot 600 μL sample to a new tube, "H2A.X." Keep the rest of the sample in −80 °C.
3. Add 2 μL anti-H3 primary antibodies to the "H3" tubes. Add 5 μL anti-H2A.X primary antibodies to the "H2A.X" tubes.
4. Rotate "H3," "H2A.X," "IgG" and "input" tubes at 4 °C overnight.
5. On the same day, Add 150 μL anti-Rabbit IgG magnetic beads to a new 1.5 mL tube (*see* **Note 13**). Put the tube on the magnetic stand for 1 min to pellet the magnetic beads. Remove the supernatants.
6. Resuspend the pellet in 1 mL 0.5 % BSA in PBS. Rotate at 4 °C overnight.

7. Put the tube on the magnetic stand for 1 min. Remove the supernatant. Resuspend the beads in 1 mL buffer D. Rotate at 4 °C for 5 min.
8. Repeat the above step twice.
9. Put the tube on the magnetic stand for 1 min. Remove the supernatant. Resuspend the beads in 150 μL buffer D.
10. Add 50 μL anti-rabbit IgG beads from **step 9** each to the "H3," "H2A.X," and "IgG" tubes. Rotate at 4 °C for 2 h.
11. Put the "H3," "H2A.X," and "IgG" tubes on a magnetic stand for 1 min. Remove the supernatants. Resuspend the beads in 1 mL buffer D. Rotate at 4 °C for 5 min.
12. Repeat above **step 5** times.
13. Put the tubes with magnetic beads on a magnetic stand for 1 min. Remove the supernatants.
14. Add 50 μL elution buffer to the tubes with magnetic beads. Put the tubes on the thermomixer and shake for 15 min at 1,400 rpm at RT.
15. Put the tubes with magnetic beads on a magnetic stand for 1 min. Collect the supernatants to new tubes.
16. Repeat **steps 14** and **15**. Collect the supernatants to the same tubes as **step 15**.
17. Perform western blot to verify the specificity of the immunoprecipitation (*see* **Note 14**).
18. After confirming the specificity of the immunoprecipitation, add 100 μL phenol–chloroform to each tube with the eluted sample from **step 16**. Add 30 μL phenol–chloroform to the input tube from **step 2**. Close the tubes tightly and shake them vigorously for 15 s.
19. Centrifuge for 5 min at 20,000 × *g*.
20. Pipette the upper aqueous phase to new tubes. Purify the samples with MinElute Reaction Cleanup Kit. Dilute the purified samples to 50 μL with Nuclease-free water.
21. Perform qRT-PCR to verify the result of the ChIP experiment. Use primers set targeting loci known to be enriched with H2A.X (Table 1) (*see* **Note 15**).
22. If both western blot and qRT-PCR verify the ChIP specificity, proceed to high-throughput sequencing to interrogate the genome-wide localization of H2A.X.

3.4 Library Construction

1. Perform end repair with End-It DNA End Repair Kit: Mix 34 μL ChIP sample from **step 20** in 3.3, 5 μL 10× End-Repair Buffer, 5 μL 2.5 mM dNTP Mix, 5 μL 10 mM ATP, 1 μL End-Repair Enzyme. Incubate at room temperature for 45 min.

Table 1
Primers targeting loci enriched with H2A.X for qRT-PCR verification

	Forward primer	Reverse primer
H2AX_1	AGGAATCCTCTTGAACAATGGGG ATCC	GCATATGCAATGTGGTTTTTCCAGCCTC
H2AX_2	TCTCTGCAGTTGCAGGAGGCCTA	AAGCTGTATGTTTGATAGTCACACTGA GGAGT
H2AX_3	GGGAGGTCATGTGACTAAGATCAA CGACT	ACTAGTGTTCAGGGTGTGAGGGTCAC
H2AX_neg	GCCACCCGGTACAGCAACTCG	TCTCGGCTGGGGTCCCGAC

Primers H2AX_1, H2AX_2, and H2AX_3 target three loci enriched with H2A.X in ESC, and primer H2AX_neg targets a locus lacking of H2A.X in ESC

2. Clean up samples with QIAquick PCR Purification Kit. Elute with 34 μL nuclease-free water.
3. Mix 34 μL DNA from **step 2**, 5 μL NEBuffer 2, 10 μL 1 mM dATP, 1 μL Klenow fragment. Incubate at 37 °C for 30 min.
4. Purify the samples with MinElute Reaction Cleanup Kit. Elute with 12 μL of nuclease-free water.
5. LigaFast and TruSeq DNA Sample Prep Kit: Mix 12 μL DNA from **step 4**, 15 μL 2× Rapid Ligation Buffer, 1 μL Adapter, 2 μL T4 DNA Ligase. Incubate at room temperature for 15 min to overnight, depending on the ligation efficiency.
6. Clean up samples with QIAquick PCR Purification Kit. Elute with 18 μL nuclease-free water. Add 2 μL 10× Gel Loading Buffer.
7. Run adapter ligated DNA samples on a 2 % E-gel EX (*see* **Note 16**).
8. Excise the large band in the range of 150–300 bp with a clean scalpel (*see* **Note 17**).
9. Extract the DNA from the agarose slices with QIAquick Gel Extraction Kit (*see* **Note 18**). Elute in 25 μL nuclease-free water.
10. Mix 20 μL extracted DNA, Phusion HF PCR Master Mix 25 μL, PCR Primer Cocktail (from TruSeq DNA Sample Prep Kit) 5 μL.
11. Run PCR with the following protocol: 30 s at 98 °C, [10 s at 98 °C, 30 s at 65 °C, 30 s at 72 °C] 10 cycles (*see* **Note 19**), 5 min at 72 °C, hold at 4 °C.
12. Clean up samples with QIAquick PCR Purification Kit. Elute with 20 μL nuclease-free water.

13. Run PCR amplified DNA samples on a 2 % E-gel EX (*see* **Note 16**).
14. Excise the large band in the range of 150–300 bp with a clean scalpel (*see* **Note 17**).
15. Extract the DNA from the agarose slices with QIAquick Gel Extraction Kit (*see* **Note 18**). Elute in 30 μL nuclease-free water.
16. Measure the sample's DNA concentration and A260/A280 by NanoDrop and Bioanalyzer 2100.
17. Perform high-throughput sequencing with Illumina HiSeq 2000.

4 Notes

1. Add fresh Protease inhibitor to buffer A before starting the experiment.
2. Triton X-100 should be supplemented at last.
3. If available, ESC deficient for the protein of interest should be used as a control, and processed in parallel with the wild-type ESC.
4. Depending on the expression level of the proteins of interest and how many ChIP experiments you want to perform with the same batch of cells, you can increase or decrease the amount of cells used. The amount of solution used in the later steps should be adjusted accordingly.
5. Before pipetting to resuspend the cells, gently tap the bottom of the tube to disrupt the cell pellet.
6. The cells become very fragile after being treated with Triton X-100. Mix the sample gently by inverting. Do not vortex to avoid breaking the cells and creating bubbles.
7. Lay the tube on top of the ice. Do not bury it in the ice.
8. Make wide-orifice tips by cutting the end of tips with a sharp blade. Using wide-orifice tips can decrease mechanical shearing and cell fragmentation.
9. It is known that chromatin in ESC obtain less heterochromatins than differentiated cells [12]. In agreement with this notion, we found that the isolation of chromatin pellets from ESC require modifications of the original N-ChIP protocol. If performing N-ChIP with differentiated cells, there is an additional step of washing the nuclei with no-salt buffer. However, due to ESC's open chromatin structure and fragile nuclei, washing with no-salt buffer will break the nuclei and hinder the later steps.
10. $CaCl_2$ serves as the catalyst for micrococcal nuclease. If the digestion condition needs to be optimized, the concentration of $CaCl_2$ can vary from 1 to 5 mM.

11. The length of the digestion is critical in the N-ChIP experiment. If multiple samples need to be processed together, either allow a 30 s gap between each sample or digest them one by one.
12. If the digestion is not complete, chromatins will not be digested to mononucleosomes, and the final ChIP signal will be of lower resolution. Thus, if the majority of the DNA is not between 150 and 200 base pairs, the micrococcal nuclease digestion conditions need to be optimized. It is suggested that before doing a large-scale ChIP experiment, one should first perform a preliminary experiment to optimize digestion conditions, including the concentration of the micrococcal nuclease, the concentration of $CaCl_2$, and the length of digestion.
13. Sometime the beads will cluster together and stuck the opening of the tips, which makes the pipetting inaccurate. Use wide-orifice tips and pipet the beads slowly to acquire the accurate amount of beads.
14. The specificities of the antibodies are highly variable among different antibodies, even among different batches of the same antibody. Thus it is important to verify the specificity of a new antibody by examining the immunoprecipitated samples by western blot.
15. Normalize the qRT-PCR results with the input DNA. Compare the H2A.X ChIP with the IgG ChIP to verify whether there is an enrichment of signal in the H2A.X ChIP.
16. If making multiple libraries at the same time, leave an empty lane between samples to avoid cross-contamination.
17. Optional: take photos of the gel before and after the gel excision to confirm a complete excision.
18. To cleanup DNA from previous enzymatic reactions, be sure to add 100 μL of isopropanol to the reaction mix after the gel slice dissolves completely in the Buffer QG.
19. A typical amplification is from 10 to 20 cycles, depending on the amount of the starting material.

References

1. Gilmour DS, Lis JT (1984) Detecting protein-DNA interactions in vivo: distribution of RNA polymerase on specific bacterial genes. PNAS U S A 81:4275–4279
2. Solomon MJ, Larsen PL, Varshavsky A (1988) Mapping protein-DNA interactions in vivo with formaldehyde: evidence that histone H4 is retained on a highly transcribed gene. Cell 53(6):937–947
3. Orlando V (2000) Mapping chromosomal proteins in vivo by formaldehyde-crosslinked-chromatin immunoprecipitation. TIBS 25: 99–104
4. Hebbes T, Thorne A, Crane-Robinson C (1988) A direct link between core histone acetylation and transcriptionally active chromatin. EMBO 7(5):1395–1402
5. O'Neill L (2003) Immunoprecipitation of native chromatin: NChIP. Methods 31:76–82
6. Turner B (2001) ChIP with native chromatin: advantages and problems relative to methods using cross-linked material. In: Mapping

protein/DNA interactions by cross-linking. INSERM, Paris
7. Bellard M, Dretzen G, Giangrande A, Ramain P (1989) Nuclease digestion of transcriptionally active chromatin. Methods Enzymol 170:317–346
8. Park PJ (2009) ChIP-seq: advantages and challenges of a maturing technology. Nat Rev Genet 10:669–680
9. Egelhofer TA, Minoda A, Klugman S, Lee K, Kolasinska-Zwierz P, Alekseyenko AA, Cheung M-S et al (2011) An assessment of histone-modification antibody quality. Nat Struct Mol Biol 18:91–93
10. Méndez J, Stillman B (2000) Association of human origin recognition complex, cdc6, and minichromosome maintenance proteins during the cell cycle: assembly of prereplication complexes in late mitosis. MCB 20(22):8602–8612
11. Illumina Inc. (2008) Preparing samples for sequencing genomic DNA
12. Meshorer E, Misteli T (2006) Chromatin in pluripotent embryonic stem cells and differentiation. Nat Rev Mol Cell Biol 7:540–546

Chapter 3

Reduced Representation Bisulfite Sequencing to Identify Global Alteration of DNA Methylation

Arvindhan Nagarajan, Christine Roden, and Narendra Wajapeyee

Abstract

Reduced representation bisulfite sequencing is a cost-effective high-throughput sequencing-based method to obtain DNA methylation status at a single-nucleotide level. DNA methylation status is determined by utilizing DNA methylation-specific restriction enzymes to selectively amplify for genomic regions that are rich in methylated DNA. Although the method is genome-wide, DNA methyl sequencing does not require the sequencing of the whole genome, hence the name "reduced representation." However, a large majority of CpG islands are covered by reduced representation bisulfite sequencing allowing for the acquisition of comprehensive information of the methylation landscape in diseases like cancer. Data generated by this approach is typically reproducible and often covers between 65 and 75 % of the whole genome.

Key words 5-Methylcytosine, RRBS, Bisulfite sequencing, CpG islands, Deep sequencing

1 Introduction

Epigenetic alterations refer to changes in the DNA that are reversible and heritable [1, 2]. These changes in DNA or chromatin often alter gene expression patterns without affecting the underlying "genetic code." Two major types of known epigenetic modifications are DNA methylation and histone modifications. Of these 5-methylcytosine (5-mC) and 5-hydroxymethyl cytosine (5hmC) are the only known covalent modifications of the mammalian DNA [2, 3]. DNA methylation plays a major role in regulation of gene expression such as in gene silencing and X-chromosome inactivation [1, 2]. DNA methylation usually takes place in the context of CpG dinucleotides in the CpG islands. Aberrant alterations in DNA methylation have been shown to be associated with many diseases such as cancer and metabolic syndrome and some other genetic diseases [1, 2]. Due to the causative role of aberrant DNA methylation in cancer, inhibitors of DNA methylation are being used as therapeutic agents in cancer [4]. The knowledge of alteration of methylation patterns in various developmental processes

Narendra Wajapeyee (ed.), *Cancer Genomics and Proteomics: Methods and Protocols*, Methods in Molecular Biology, vol. 1176, DOI 10.1007/978-1-4939-0992-6_3, © Springer Science+Business Media New York 2014

and diseases is essential to understand the role of DNA methylation and for designing specific therapeutic interventions.

There are several methods that are being used to study DNA methylation patterns of specific gene loci or at the whole genome scale. Conventional DNA sequencing methods cannot differentiate 5-methylcytosine from cytosine. Therefore, selective enrichment of methylated genomic DNA needs to be performed before sequencing. 5-Methylcytosine can be physically identified by three methods: (1) an antibody against 5-methylcytosine, (2) the use of methyl-binding domains of various proteins that specifically bind to 5-methylcytosine, and (3) the use of certain restriction enzymes that specifically cut unmethylated DNA recognition sequences. These three methods are employed in Me-DIP sequencing, Me-CAP sequencing, and methylation-sensitive digestion and sequencing, respectively. These methods allow the selective enrichment of methylated DNA and sequencing of only DNA-methylated regions. However, none of these methods provide DNA methylation information at the single-nucleotide level. Bisulfite sequencing on the other hand uses an initial step of bisulfite treatment, which converts cytosine to uracil and in turn can then be differentiated by conventional DNA sequencing and thus provides DNA methylation information at a single-nucleotide level. By combining bisulfite sequencing to the whole-genome sequencing DNA methylation information at a single nucleotide can be achieved for the entire genome. However, similar to other all whole-genome sequencing methods it is time consuming and expansive. Moreover, because DNA methylation is typically restricted to CpG islands this method is also very inefficient. RRBS overcomes this problem by performing bisulfite sequencing only on DNA methylation-rich sequences. The initial step of enrichment of methylation-dense region is achieved by digestion with methyl-specific restriction enzyme, MspI, followed by adapter ligation and bisulfite conversion of these regions. Following this the DNA is PCR amplified and sequenced. For a detailed comparison of RRBS with other techniques used to study methylation please refer to Bock et al. [5].

The protocol presented in this chapter is modified gel-free multiplexed reduced representation bisulfite sequencing for large-scale DNA methylation profiling [6]. The conventional RRBS uses gel separation to isolate fragment sizes between 40 and 350 bp [7–9]. The current protocol achieves the same fragment size selection as conventional RRBS by using solid-phase reversible immobilization (SPRI) to purify DNA fragments after adapter ligation, which removes fragments <40 bp followed by cluster amplification on an Illumina flow cell, which select against large fragments. Because of cost savings achieved by using gel extraction over SPRI cleaning, when handling small sample numbers the protocol is provided as an option (*see* **Note 2**).

2 Materials

2.1 Reagents

1. Quant-iT DNA Broad Range assay kit.
2. DNA isolation kit.
3. Gel purification kit.
4. 10 mM Tris–HCl, 0.1 mM EDTA, PH 8.0.
5. 96-Well PCR plate.
6. CpG Methylated NIH 3T3 mouse genomic DNA.
7. MspI restriction enzyme.
8. Adhesive tape sheet.
9. Klenow fragment.
10. 10 mM dATP, 1 mM dCTP, and 1 mM dGTP.
11. SPRI AMPure XP beads.
12. DynaMagTM-96 Side magnet.
13. 70 % Ethanol.
14. 20 μl of EB buffer.
15. T4 ligase.
16. Illumina TruSeq adapter (containing 5 mC) (Illumina, Dedham, MA, USA, catalogue number PE-940-2001).
17. 1.5 ml Microfuge tubes.
18. 20 % Polyethylene glycol (8,000-g/mol).
19. 2.5 M NaCl.
20. PfuTurbo Cx hotstart DNA polymerase.
21. 0.3 μM of TruSeq primers (forward primer, 5′-AATGATAC GGCGACCACCGAGAT-3′; reverse primer, 5′-CAAGCAGA AGACGGCATACGA-3′).
22. Qubit dsDNA HS assay kit.
23. 4–20 % Polyacrylamide TBE gel.

2.2 Equipment

1. Plate centrifuge.
2. Incubator.
3. Thermocycler.
4. 8-Channel pipettes.
5. 12-Channel pipettes.
6. Fume hood.
7. Qubit fluorometer.
8. Illumina Hiseq 2000.

3 Methods

3.1 Genomic DNA Purification

1. Purify genomic DNA from cells or tissue samples using DNA isolation kit as per the manufacturer's protocol. Briefly resuspend cells (2×10^3 to 5×10^6) or sliced tissues in 30–400 μl lysis buffer. Vortex briefly to suspend cell pellets.
2. Incubate the lysis reaction at 55 °C for 3 h to digest cellular proteins and RNA.
3. Add an equal volume of 1× PBS buffer to the lysis reaction, mix thoroughly by vortexing, and spin down briefly. Next, add an equal volume of 100 % ethanol. Mix well, and spin down briefly.
4. Isolate DNA by using a spin column supplied in the genomic DNA kit.
5. Elute the genomic DNA from the column with 30–100 μl of DNA elution buffer (10 mM Tris–HCl (pH 9.0), 0.1 mM EDTA) from the genomic DNA kit.
6. Determine the genomic DNA concentration with a Qubit fluorometer and Quant-iT DNA assay kit, broad range as per the manufacturer's instructions. Dilute the genomic DNA to 20 ng/μl using TE. 100 ng of genomic DNA is used for each sample.

3.2 MspI Digestion

1. Digest 100 ng (5 μl) of genomic DNA in each well of a 96-well PCR plate.
2. The reaction is carried out in a 30 μl volume with 10 U of MspI as indicated in Table 1. A master mix is prepared according to the number of samples based on the following reaction setup.
3. Add 25 μl of MspI master mix per well to the plate, spin down, vortex, and spin for 30 s at 2,000 × *g*. Incubate overnight at 37 °C.

3.3 Gap Filling and A-Tailing

1. Gap filling and A-tailing are carried out with Klenow fragment and dNTP mix in the same plate containing the MspI-digested genomic DNA. A master mix is prepared according to the number of samples based on the following reaction setup (Table 2).

Table 1
MspI digestion

Reagents	Final concentration	Final volume (μl)
10× NEB buffer 2	1×	3
Genomic DNA (20 ng/μl)		5
MspI	10 U	1
H_2O		21
Total		30

Table 2
Gap filling and A-tailing

Reagents	Final concentration	Final volume (μl)
MspI digestion mix		30
dNTP mixture (10 mM dATP, 1 mM dGTP, 1 mM dCTP)	0.5 mM dATP, 0.05 mM dGTP, 0.05 mM dCTP	1
Klenow fragment	5 U	1
Total		32

Table 3
Multiplexed adapter ligation

Reagents	Final concentration	Final volume (μl)
SPRI beads in EB		20
10× T4 ligation buffer	1×	3
T4 ligase		1
Illumina TruSeq adapter (450 nM)	45 nM	2
H_2O		4
Total		30

2. Add 2 μl of master mix per well to the plate. Seal the plate, spin down, vortex, and spin for 30 s at 2,000 × *g*.
3. Incubate the plate in a thermocycler at 30 °C for 20 min followed by 37 °C for 20 min.

For gel-based size exclusion (*see* **Note 1**):

3.4 First SPRI Cleanup

1. Add 64 μl SPRI beads for 32 μl sample, and pipette up and down five times to mix.
2. Incubate the samples at room temperature for 30 min.
3. Place on a DynaMagTM-96 Side magnet for 5 min, and remove supernatant being careful not to disturb the beads.
4. Wash the beads twice in 70 % ethanol. Wait for 5 min, and then remove the remaining ethanol. Dry the beads for 10 min in a fume hood.
5. Add 20 μl of EB buffer, then seal, vortex, and spin down.

3.5 Multiplexed Adapter Ligation

1. The multiplexed adapter ligation reaction is carried out in a 30 μl reaction. A master mix is prepared according to the number of samples based on the following reaction setup (Table 3).
2. Add 20 μl of master mix to each of the well containing the SPRI beads.

Table 4
PCR reaction

Reagents	Final concentration	Final volume (μl)
PfuTurbo Cx hotstart DNA polymerase buffer (10×)	1×	20
dNTP (25 mM each, 100 mM)	0.25 mM each	2
Standard/paired-end Illumina adapters (2.5 μM for each)	0.3 μM	24
PfuTurbo Cx hotstart DNA polymerase (2.5 U/μl)	10 U	4
DNA templates		24
H_2O		126
Total		200

3. Incubate in a thermocycler at 16 °C overnight (unheated lid).
4. Remove the plate from the thermocycler, and resuspend the beads.
5. Deactivate the enzyme at 65 °C for 20 min in a thermocycler.

3.6 Library Pooling and Bisulfite Conversion

1. Pool the 12 samples with unique barcodes into 1.5 ml tube.
2. Add 720 μl of 20 % polyethylene glycol and 2.5 M NaCl. Incubate at room temperature for 30 min.
3. Place samples into DynaMag Magnet for 5 min.
4. Wash the beads with 70 % ethanol. Remove ethanol, and place samples in fume hood to dry for 30–50 min or until cracks appear.
5. Elute DNA from beads by adding 25 μl of EB buffer, vortex for 20 s, and centrifuge briefly.
6. Place samples on magnet, and transfer eluant to fresh 1.5 ml tubes. Save 3 μl of each sample for ligation efficiency test.
7. Bisulfate convert the remaining 20 μl of DNA twice using a large-scale PCR reaction (200 μl).
8. Perform large-scale PCR reaction as follows (Table 4).

For gel-based size exclusion (*see* **Notes 1** and **2**):

3.7 Final SPRI Bead Cleanup

1. Transfer PCR product to 1.5 ml tubes.
2. Clean with 240 μl SPRI beads into a 200 μl library pool.
3. Elute DNA in 40 μl of EB buffer. Repeat SPRI cleanup. Elute with 40 μl EB buffer.
4. Quantify the DNA concentration using Qubit dsDNA HS assay kit.
5. Check the quality of the sample using 4–20 % polyacrylamide TBE gel.

3.8 Sequencing and Alignment

1. Perform dark sequencing (imaging and cluster localization) using Illumina Hiseq 2000. (No images are recorded during the first three sequencing cycles.)
2. Remove sequences of adapters and barcodes. For suggested data analysis pipeline *see* **Note 3**.
3. Align the 29 bp reads following the barcode to the hg19 genome using MAQ software.
4. Identify CpG methylation marks by observing the bisulfite transformation in the read as opposed to the genome sequence.

For available bioinformatics algorithms for analysis of bisulfite conversion *see* **Note 4**.

4 Notes

1. If a gel-based size exclusion is to be done instead of SPRI cleanup instead of Subheading 3.4 purify DNA using phenol:chloroform extraction and perform Subheading 3.6. Skip Subheading 3.7, and instead perform gel extraction after Subheading 3.6 as explained in **Note 2**. For phenol:chloroform extraction bring the volume of sample to 200 μl with TE and add equal volume of phenol:chloroform:isoamyl alcohol (25:24:1). Mix, spin at maximum speed for 5 min, and transfer supernatant to a fresh tube. Precipitate DNA with ethanol and glycogen, and wash pellet with 70 % ethanol.
2. Gel-based size selection of the adapter-ligated DNA fragments: Prepare a 3 % Nusieve 3:1 agarose 0.5× TBE gel without ethidium bromide. Add 1 μl (50 ng) carrier DNA to each sample. Load DNA samples into the gel wells with sufficient space between adjacent samples along with marker. Excise the DNA ladder lanes from the gel and stain with SYBR Green I. With the DNA ladder as the guide cut out gel between 160 and 340 bp (these sizes correspond to 40 and 220 bp restriction fragments, as they run slower than the marker). The DNA is extracted from the gel using gel purification kit.
3. Suggested data analysis pipeline:
   Generate FASTQ files using the latest version of CASAVA. Transfer FASTQ files and latest version of the human genome (FASTA formatted) to a high-performance computing cluster. Download and install the latest versions of BWA, Picard, JAVA, and Bio::DB::Sam Perl. Align the data with BWA so that it is compatible with Picard tools and GATK, http://bio-bwa.sourceforge.net/bwa.shtml. Be sure to include an appropriate read group header. Convert the SAM file to a BAM file using Picard.

4. Available bioinformatics algorithms for analysis of bisulfite conversion:
   (a) BSMAP: The algorithm is capable of identifying methylation sites on asymmetric bases after bisulfite treatment on a read length of up to 144 nucleotides with 15 mismatches. Supports SAM format input (https://sites.google.com/a/brown.edu/bioinformatics-in-biomed/bsmap-for-methylation).
   (b) RMAP-bs: For FASTA-formatted files this algorithm. is capable of identifying C-to-T changes in comparison to the reference genome, http://rulai.cshl.edu/rmap/.
   (c) BS Seeker: For SAM- or BAM-formatted data. Requires both Bowtie and Python to align to the reference genome. Useful for identifying CG, CHG, and CHH reads, http://pellegrini.mcdb.ucla.edu/BS_Seeker/USAGE.html.
   (d) Bismark: Determines the unique best alignment from four parallel alignment processes. Methylation state of positions involving cytosines (C → T) and guanines (G → A) is determined by comparisons to the reference genome. User friendly for biologists, http://www.ncbi.nlm.nih.gov/pmc/articles/PMC3102221/.
   (e) CgiHunter: Useful in identifying regions of the genome with a greater proportion of GC content and CpG dinucleotide sequences and coordinating these locations with methylation status, http://cgihunter.bioinf.mpi-inf.mpg.de/.

Acknowledgements

N.W. is a translational scholar of the Sidney Kimmel Foundation for Cancer Research and is supported by the Young Investigator Awards from National Lung Cancer Partnership, Uniting Against Lung Cancer, and International Association for the Study of Lung Cancer, Melanoma Research Alliance, and Melanoma Research Foundation.

References

1. Deaton AM, Bird A (2011) CpG islands and the regulation of transcription. Genes Dev 25(10):1010–1022, PubMed PMID: 21576262. Pubmed Central PMCID: 3093116. Epub 2011/05/18. eng
2. Jones PA, Baylin SB (2007) The epigenomics of cancer. Cell 128(4):683–692, PubMed PMID: 17320506. Epub 2007/02/27. eng
3. Wu SC, Zhang Y (2010) Active DNA demethylation: many roads lead to Rome. Nat Rev Mol Cell Biol 11(9):607–620, PubMed PMID: 20683471. Pubmed Central PMCID: 3711520. Epub 2010/08/05. eng
4. Gnyszka A, Jastrzebski Z, Flis S (2013) DNA methyltransferase inhibitors and their emerging role in epigenetic therapy of cancer. Anticancer

Res 33(8):2989–2996, PubMed PMID: 23898051. Epub 2013/07/31. eng

5. Bock C, Tomazou EM, Brinkman AB et al (2010) Quantitative comparison of genome-wide DNA methylation mapping technologies. Nat Biotechnol 28(10):1106–1114, PubMed PMID: 20852634. Pubmed Central PMCID: 3066564. Epub 2010/09/21. eng
6. Boyle P, Clement K, Gu H et al (2012) Gel-free multiplexed reduced representation bisulfite sequencing for large-scale DNA methylation profiling. Genome Biol 13(10):R92, PubMed PMID: 23034176. Pubmed Central PMCID: 3491420. Epub 2012/10/05. Eng
7. Meissner A, Gnirke A, Bell GW et al (2005) Reduced representation bisulfite sequencing for comparative high-resolution DNA methylation analysis. Nucleic Acids Res 33(18):5868–5877, PubMed PMID: 16224102. Pubmed Central PMCID: 1258174. Epub 2005/10/15. eng
8. Gu H, Bock C, Mikkelsen TS et al (2010) Genome-scale DNA methylation mapping of clinical samples at single-nucleotide resolution. Nat Methods 7(2):133–136, PubMed PMID: 20062050. Pubmed Central PMCID: 2860480. Epub 2010/01/12. eng
9. Gu H, Smith ZD, Bock C et al (2011) Preparation of reduced representation bisulfite sequencing libraries for genome-scale DNA methylation profiling. Nat Protoc 6(4):468–481, PubMed PMID: 21412275. Epub 2011/03/18. eng

Chapter 4

A High-Throughput MicroRNA Expression Profiling System

Yanwen Guo, Stephen Mastriano, and Jun Lu

Abstract

As small noncoding RNAs, microRNAs (miRNAs) regulate diverse biological functions, including physiological and pathological processes. The expression and deregulation of miRNA levels contain rich information with diagnostic and prognostic relevance and can reflect pharmacological responses. The increasing interest in miRNA-related research demands global miRNA expression profiling on large numbers of samples. We describe here a robust protocol that supports high-throughput sample labeling and detection on hundreds of samples simultaneously. This method employs 96-well-based miRNA capturing from total RNA samples and on-site biochemical reactions, coupled with bead-based detection in 96-well format for hundreds of miRNAs per sample. With low-cost, high-throughput, high detection specificity, and flexibility to profile both small and large numbers of samples, this protocol can be adapted in a wide range of laboratory settings.

Key words Noncoding RNAs, miRNAs, High-throughput miRNA labeling, miRNA capture

1 Introduction

MicroRNAs (miRNAs) are short, ~22 nucleotide-long noncoding RNAs that posttranscriptionally regulate protein-coding gene expression [1]. Mounting evidence shows that miRNAs regulate a large number of mammalian biological processes, ranging from physiological responses to pathological mechanisms [2]. For example, deregulation of miRNAs regulates cancer initiation and progression [3, 4]. In addition to their functional importance, miRNA expression has been found to richly reflect diagnostic and prognostic information in human patients and reflect pharmacological responses [4, 5]. Taken together, the increasing interest in miRNAs leads to the need to profile miRNA expression for a large amount of samples. Contrasting to this need, however, most miRNA profiling techniques are both expensive and not compatible with high-throughput sample processing. We describe here an inexpensive miRNA expression profiling method that

Authors Yanwen Guo and Stephen Mastriano contributed equally to this work.

Narendra Wajapeyee (ed.), *Cancer Genomics and Proteomics: Methods and Protocols*, Methods in Molecular Biology, vol. 1176, DOI 10.1007/978-1-4939-0992-6_4, © Springer Science+Business Media New York 2014

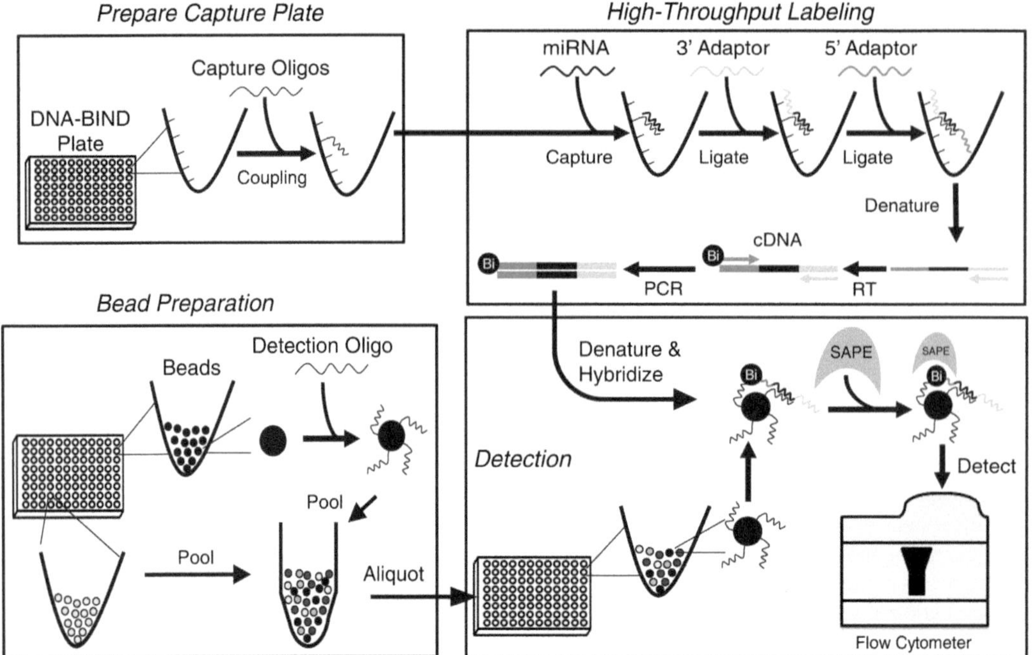

Fig. 1 Schematic of the high-throughput miRNA profiling. The major steps of the protocol are shown. *Bi* biotin; *SAPE* streptavidin–phycoerythrin

supports both high-throughput sample processing and high-throughput detection of hundreds of miRNAs simultaneously. In addition to meeting the need to study regulation and deregulation of miRNAs in various biological contexts on a small scale, this method makes it possible to quickly conduct large-scale studies in which hundreds or thousands of samples are required, such as large collections of cancer-related specimens, and cellular pharmacological responses from thousands of chemicals.

Although the importance of miRNAs is well known, it is less well appreciated that cellular miRNAs exist in multiple isoforms. The biogenesis pathway of mature miRNAs determines that newly produced mature miRNAs have a 5′ phosphate group and a 3′ hydroxyl group. Despite the fact that groups such as 5′ phosphate help miRNA interact with Argonaut protein [6], it is possible that some cellular miRNAs exist in dephosphorylated forms or without 3′ free hydroxyl group. It is thus important to emphasize that different methods of measuring miRNA expression have different specificity toward different miRNA isoforms. For example, many commercial miRNA profiling and quantitative RT-PCR methods cannot differentiate between 5′ phosphorylated and unphosphorylated forms of miRNAs (e.g., [7]). The method presented here detects the functional mature miRNA isoforms with both 5′ phosphate and 3′ hydroxyl groups.

We describe here a protocol for both high-throughput sample labeling and sample detection for miRNA profiling (Fig. 1). Using this protocol, it is easily achievable to obtain miRNA expression profiles for over 300 samples in ~3 days by a single person.

In addition to the throughput and specificity toward functional mature miRNAs, this system is also convenient to implement, has low per-sample cost, and has high detection specificity [5]. Specifically, extending on previous methods [5, 8], this high-throughput labeling technique starts with total RNAs (easily obtainable by methods such as TRIzol-based purification) and utilizes hybridization-based miRNA capturing in a 96-well format, so that mature miRNAs can be enriched by binding to complementary probes. Subsequent biochemical reactions are performed in the same capture plate, involving ligating mature miRNAs with 3′ and 5′ adaptors consecutively and reverse transcribing from a primer complementary to the 3′ adaptor. Lastly, PCR-based labeling is then performed with a common pair of primers matching 5′ and 3′ adaptors, with the 5′ adaptor-matching primer containing a biotin label. Majority of the above labeling steps occur on the same capture plate, which significantly improves the throughput of sample labeling.

In addition to high-throughput sample labeling, we also describe here a protocol for high-throughput miRNA profiling detection method that supports batch sample detection in 96-well plates (Fig. 1). This method involves hybridization of labeled biotinylated miRNAs to colored beads with covalently bound detection probes (with up to 500 colors to choose from), followed by binding of the fluorescent streptavidin–phycoerythrin to biotin and then by flow cytometry to identify the specific bead color (reflecting miRNA identity) and quantify phycoerythrin intensity on the bead (reflecting the level of hybridized miRNAs). Notably, the hybridization of miRNA probes to colored beads mimics liquid-phase hybridization kinetics, leading to enhanced hybridization performance and specificity compared to hybridization on a 2-dimensional surface [5]. Specifically, the detection beads are created by covalently coupling 5′-amino-modified oligonucleotide-capture probes (complementary to miRNAs) to carboxylated 5 μm polystyrene beads, with one miRNA coupled to beads with a specific color combination. Bead color is determined by two fluorescent dyes at variable ratios (up to 500 combinations) embedded in the beads and can be easily differentiated with flow cytometry.

Overall, this protocol requires ~2 days to perform high-throughput miRNA labeling and ~1 day to perform high-throughput miRNA detection for hundreds of samples. This protocol has been applied to thousands of samples with robust performance (e.g., [5, 8–13]). Note that we only describe the bench-level protocol to obtain miRNA profiling data. Another important component of any profiling experiment is computational analysis after the data acquisition, which is often project specific and hence not covered here. Examples of analyses of data obtained with this method can be found in publications [5, 8, 9, 12–14]. In addition, as with all profiling techniques, follow-up experiments are often required to validate profiling results. Northern blots and quantitative RT-PCR can be used.

2 Materials

Prepare all reagents in RNase-free water (e.g., Sigma W4502) and store at room temperature, unless indicated otherwise. All chemicals are from Sigma, unless specified. For buffer preparation, dilution from premade nuclease-free stock solutions from commercial sources is recommended to avoid nuclease contamination. Precautions against PCR-carryover contamination (such as using filter tips) and against RNase contamination (such as cleaning pipettes and working on a clean bench) are recommended during the miRNA labeling procedure.

The protocol below utilizes multichannel pipettes (at least three multichannels with capacities of 10, 50, and 200 μl each) that allow operating row by row or column by column. Automatic liquid-handling robots can also be used to increase the operational efficiency; we have used a Cybio liquid handler successfully with this method.

2.1 Preparation of miRNA-Capture Plate

1. DNA-BIND Thermowell-M PCR Plate (Corning Costar 6573). The active sites in the wells of the plate react with primary amine groups. (Note: Avoid primary amine-containing reagents, such as Tris, during the coupling procedure, except for the capture oligos.)
2. Plate-coupling buffer: 500 mM Na_2HPO_4, pH 8.5, 1 mM EDTA.
3. Plate-coupling wash buffer: 100 mM Tris–HCl pH 8.0, 150 mM NaCl, and 1 mM EDTA.
4. TE wash/storage buffer: 10 mM Tris–HCl pH 8.0 and 1 mM EDTA.
5. Capture oligonucleotide mix: These are 5′-amino oligonucleotides that are complementary to mature miRNAs. Such oligonucleotides are each prepared in 50 μM stocks (resuspended in double-deionized water) and then combined together into a mixture containing 50 μM of total oligonucleotides. The sequences for the oligonucleotides that we used for miRNA capture and detection have been deposited into the GEO database (platform GPL16690).
6. Microseal-A Film (Bio-Rad MSA-5001): Note that this film is semi-adhesive and can be reused for the same plate if not contaminated. To seal with a Microseal-A Film, it is recommended to use a plate roller (e.g., Bio-Rad MSR-0001).

2.2 High-Throughput miRNA Labeling

1. 4× Hybridization buffer: 4 M NaCl, 200 mM Tris–HCl pH 8.0, and 4 mM EDTA. Aliquot, and store at −20 °C.
2. RNAsecure (Invitrogen AM7005): Aliquot, and store at −20 °C.

3. Artificial small RNA mix: Prepare a mixture of the following three RNA oligonucleotides (ordered from Dharmacon; resuspend in 100 μM stocks, and store in −80 °C) at 30 nM pCAGUCAGUCAGUCAGUCAGUCAG, 30 nM pGACCUCCAUGUAAACGUACAA, and 5 nM pUUGCAGAUAACUGGUACAAG. Aliquot, and store at −80 °C. The purpose of these artificial small RNAs is to control the efficiency of the labeling procedure.
4. 2× SSC: Dilute from 20× SSC (American Bioanalytical AB13156-01000).
5. 5× Ligation buffer: 250 mM Hepes–NaOH pH 8.3, 50 mM $MgCl_2$, 16.5 mM DTT, 50 μg/ml acetylated BSA (Invitrogen AM2614), and 41.5 % glycerol. Store at −20 °C.
6. T4 RNA Ligase (Fermentas EL0021): Store at −20 °C.
7. Ligation adaptor: 3′ Adaptor pUUUdAdAdCdCdGdCdGdAdAdTdTdCdCdAdGidT; 5′ adaptor dAdCdGdGdAdAdTdTdCdCdTdCdAdCdTAAA (ordered from Dharmacon). DNA bases are labeled as dN, whereas RNA bases are N. Resuspend in RNase-free water at 100 μM, aliquot, and store at −20 °C.
8. First-strand synthesis kit, including 5× reverse transcription buffer, 10 mM dNTP, 0.1 M DTT (Invitrogen 11904-018). Store at −20 °C.
9. MMLV-RT enzyme (Invitrogen 28025-021): Store at −20 °C.
10. Primer: PCR-R1:TACTGGAATTCGCGGTTA; PCR-F1: 5′-biotin-CAACGGAATTCCTCACTAAA. Resuspend in RNase-free water at 100 μM, aliquot, and store at −20 °C.
11. Taq DNA Polymerase with 10× ThermoPol buffer (NEB M0267L). Store at −20 °C.
12. dUTP/dATP/dCTP/dTTP mix (10 mM each, referred to as "dUTP mix," Promega U1335). Aliquot, and store at −20 °C.
13. 96-Well PCR plates.
14. A centrifuge capable of spinning 96-well plates.
15. A PCR machine that accepts 96-well plates.

2.3 Preparation of Detection Beads

1. Luminex MicroPlex or MagPlex Microspheres (Luminex, *see* **Note 1**): Concentration of beads is normally ~12.5 million/ml from the manufacturer. The active sites on the beads react with primary amine groups in the presence of EDC. (Note: Avoid primary amine-containing reagents, such as Tris, during the coupling procedure, except for the detection oligos.)
2. Detection oligonucleotides: These are the same oligonucleotides described in reagent 5, Subheading 2.1. Stock concentration is 50 μM each.
3. 96-Well PCR plates.

4. 0.1 M 2[N-morpholino] ethanesulfonic acid (MES; prepare with NaOH to reach pH 4.5), store at 4 °C.
5. EDC (Pierce 77149): Store at −20 °C.
6. 0.1 % SDS in double-deionized water.
7. 0.02 % Tween-20 in double-deionized water.

2.4 Hybridization and Detection

1. Hybridization control mix: 100 nM/5Biosg/CAACGGAATTCCTCACTAAACCCTGGACAGGCGAGGAATACAGTTTAACCGCGAATTCCAGTA), 30 nM/5Biosg/CAACGGAATTCCTCACTAAATGACTCTCAGCGAGCCTCAATGCTTTAACCGCGAATTCCAGTA, and 5 nM/5Biosg/CAACGGAATTCCTCACTAAACTGCGGGAGCCGATTTCATCTTTAACCGCGAATTCCAGTA. Aliquot, and store at −20 °C. The purpose of these hybridization controls is to control the efficiency of hybridization and detection.
2. 1.5× TMAC buffer: For each 50 ml, add 45 ml 5 M TMAC (Sigma T3411-1L), 0.375 ml 20 % sarkosyl, 3.75 ml 1 M Tris–Cl pH 8.0, 0.6 ml 0.5 M EDTA, and 0.275 ml water. Dilute 1.5× TMAC buffer to 1× TMAC buffer using water.
3. Streptavidin R–phycoerythrin (SAPE) conjugate (Invitrogen S-866), store at 4 degree in the dark.
4. Luminex FlexMap 3D machine (Luminex, *see* **Note 2**).

3 Methods

3.1 Preparation of miRNA-Capture Plate

1. Freshly dilute capture oligonucleotide mix to a final concentration of 5 μM (total oligo concentration) in plate-coupling buffer.
2. Add 20 μl/well of the diluted capture oligonucleotide mix into desired wells of one or more DNA-BIND Thermowell-M PCR plates using a multichannel pipette. When adding the capture oligonucleotide mix, make sure that the tips go to the bottom of the wells and liquid is dispensed slowly so as to stay at the bottom of the well (to avoid coupling to the wall of the plate above 20 μl of volume). Cover plate with a Microseal-A Film (or plastic wrap). Let the coupling reaction go for 1–1.5 h.
3. Using a multichannel pipette, remove the capture oligonucleotide-containing buffer, and add 180 μl/well of plate-coupling wash buffer. Let the plate stay in room temperature for 15 min or more to quench the active sites on the plate.
4. Wash the plate once with plate coupling buffer (with ~180 μl/well). Then wash twice with TE wash/storage buffer (with ~180 μl/well). After the last wash, remove TE wash/storage buffer to ~100 μl remaining in each well. Briefly spin to collect volume.

5. Coupled miRNA-capture plate can be used immediately or can be stored at 4 °C for at least 1 month without loss of quality. If storage is desired, make sure that the plate is properly covered (such as with a Microseal-A Film) to avoid liquid evaporation.

3.2 High-Throughput miRNA Labeling

3.2.1 miRNA Capture

1. Total RNA samples can be prepared using TRIzol (Invitrogen 15596-026) and isopropanol precipitation, following the manufacturer's protocol. Use 250 ng to 2 μg of total RNA per sample for labeling (*see* **Note 3**).
2. Add total RNA samples to a 96-well PCR plate (referred to as the "sample preparation plate") (*see* **Note 4**). Add RNase-free water so that total volume is equal among all samples. The maximum volume for RNA plus water is 14 μl.
3. Prepare a master mix for miRNA capture. For each sample well, use 5 μl 4× hybridization buffer and 0.8 μl RNAsecure. In addition, add 0.01 μl artificial small RNA mix per 1 μg total RNA per well (pre-dilute the artificial small RNA mix for accurate pipetting if necessary). Finally, RNase-free water can be added, so that the total volume of miRNA capture reaction is 20 μl/well, which includes hybridization buffer, RNAsecure, artificial small RNA mix, total RNA, and water.
4. Using a multichannel pipette, add a suitable amount of master mix to wells on the sample preparation plate, so that each component is as specified in **step 3** above. Mix by pipetting up and down a few times. Briefly spin this sample preparation plate to collect volume at the bottom of wells.
5. Take a coupled miRNA-capture plate prepared in Subheading 3.1 (referred to as "reaction plate") and briefly spin to collect TE wash/storage buffer at the bottom of wells. Use a multichannel pipette to partially remove the TE wash/storage buffer so that ~30–50 μl is remaining. Briefly spin the plate again, and then completely remove all the remaining TE wash/storage buffer in the wells.
6. Completely transfer the samples in the sample preparation plate into the reaction plate. This step should be done immediately after the previous step. In general, avoid drying up the wells of the reaction plate during the whole procedure. Seal the plate with a Microseal-A Film. Briefly spin to collect volume, and incubate on a PCR machine with the following program: 80 °C 4 min; touchdown step: from 80 °C for 1 min per cycle, 20 cycles, with a decrease of 1 °C per cycle; touchdown step: from 60 °C for 2 min per cycle, 30 cycles, with a decrease of 1 °C per cycle; and hold at 30 °C.
7. With a multichannel pipette, completely remove the miRNA capture mixture in the reaction plate. Wash wells three times with ~180 μl 2× SSC. After the last wash, retain ~30–50 μl of

2× SSC and cover the plate with a Microseal-A Film. Briefly spin down the plate (*see* **Note 5**).

3.2.2 Adaptor Ligation

1. Prepare 3′ ligation master reaction mix at the end of the hybridization reaction (*see* **Note 6**). The reaction mixture for each well is 3.53 μl 5× ligation buffer, 1.5 μl 100 μM 3′ adaptor, 0.94 μl 10 mM ATP, 1.18 μl T4 RNA ligase, and 12.85 μl RNase-free water (20 μl in total). The reaction mix should be prepared on ice.
2. For each well, remove the remaining liquid in the reaction plate and quickly add 20 μl 3′ ligation reaction mixture. Cover the plate with a Microseal-A film, and briefly spin down the plate. React at room temperature for 5 h to overnight.
3. After the ligation reaction, remove the reaction mixture and wash with 2× SSC five times as described in **step** 7 of Subheading 3.2.1 above.
4. Prepare 5′ ligation reaction mixture at the end of 3′ ligation reaction. The reaction mixture is the same as 3′ ligation, except that the 5′ adaptor substitutes for the 3′ adaptor.
5. Remove the remaining 2× SSC in the reaction plate, and add 20 μl of 5′ ligation reaction mixture into each well. Cover the plate with a Microseal-A film, and briefly spin down the plate. React at room temperature for 5 h to overnight.
6. After the ligation reaction, remove the reaction mixture and wash with 2× SSC five times as described in **step 3** of Subheading 3.2.2 above.

3.2.3 Reverse Transcription (RT) and PCR Labeling

1. Prepare master mix for RT mix 1 and RT mix 2 at the end of the ligation reaction. For RT mix 1, each well requires 1.6 μl PCR-R1 primer with 18.4 μl RNase-free water (20 μl in total). For RT mix 2, each well contains 8 μl 5× RT buffer, 2.24 μl 10 mM dNTP, 6.4 μl 0.1 M DTT, 2 μl MMLV-RT enzyme, and 3.36 μl water (22 μl in total).
2. Spin down the reaction plate, and remove the liquid. Using a multichannel pipette, add 20 μl/well of RT mix 1. Briefly spin again. Put the plate on the PCR machine. Heat to 80 °C for 5 min. Place the plate on ice immediately after the incubation. Cool the plate down.
3. Add 22 μl/well of RT mix 2 into each well with a multichannel pipette. Briefly spin again. Put the plate on the PCR machine, and run the following program for RT: 42 °C for 1 h and 95 °C for 5 min to inactivate. Immediately place the plate on ice to cool down.
4. Spin briefly to collect volume. Transfer the RT products to a new PCR plate. The RT products can be used in PCR labeling immediately or stored in −80 °C for many months.

5. PCR-label the RT products by using a sense primer containing a biotin label and an unlabeled antisense primer. For each RT product, set up the PCR by using 5 μl 10× Thermo PCR Buffer, 1 μl 10 mM dUTP mix, 0.5 μl PCR-F1 primer, 0.5 μl PCR-R1 primer, 0.5 μl NEB Taq polymerase, 4 μl RT product, and 38.5 μl water (50 μl in total for each well). Run PCR for 20–30 cycles (depending on the input RNA amount). The exact number of PCR cycles can be easily optimized with a pilot experiment with any given PCR machine. PCR program: 95 °C 2 min; 95 °C 30 s, 50 °C 30 s, and 72 °C 40 s, multiple cycles; and 72 °C 1 min.

3.3 Preparation of Detection Beads

1. Design the detection system so that each bead type with a distinct color combination corresponds to one miRNA detection probe for any given set of beads (referred to as "bead set") to be detected simultaneously (*see* **Note 7**).
2. Aliquot Luminex carboxylated MicroPlex or MagPlex microspheres (referred to as "beads") into 96-well plates. For each bead color (defined by different "detection regions"), aliquot 200 μl into one well on the plate. Note: The beads contain fluorophores, so it is best to avoid photobleaching by shielding from light.
3. Spin the plate at 1,600 × *g* for 3 min. Remove most of the liquid (it is fine to leave ~10 μl in each well).
4. Resuspend beads in each well with 75 μl 0.1 M MES buffer, add 2 μl 50 μM probes, and use a multichannel pipette to mix the buffer and beads by pipetting up and down.
5. Freshly make 10 mg/ml EDC in double-deionized water (not RNase-free water), quickly add 2.5 μl into each well, and mix with a multichannel pipette. React for 30 min at room temperature, shielded from light.
6. Repeat **step 5** above with freshly made EDC solution.
7. Quench the reaction by adding 0.5 M Tris–HCl pH 8.0 and 5 mM EDTA for 30 min to overnight.
8. Spin down the beads at 1,600 × *g* for 3 min. Remove the liquid in wells.
9. Resuspend beads by adding 100 μl TE wash/storage buffer into each well. Transfer all beads that belong to a bead set (for simultaneous detection, *see* **Note 7**) into a 15 ml conical tube.
10. Spin down the beads at 1,600 × *g* for 3 min. (Note that in buffers without detergent, some beads may adhere to the side of the tube wall. This is fine, but be careful not to disturb these beads when removing liquid.) Remove buffer.
11. Wash beads consecutively with 10 ml 0.02 % Tween-20, 0.1 % SDS, and TE wash/storage buffer.

12. Spin down beads, and remove the remaining liquid. Resuspend beads in each bead set in 5.95 ml TE wash/storage buffer. Beads are ready and can be stored at 4 °C for over 2 years.

3.4 Hybridization and Detection

1. Labeled PCR products from Subheading 3.2 are used in hybridization onto detection beads and then detected. For every sample with 50 μl PCR volume, add 0.05 μl of hybridization control mix and 5 μl 50 mM EDTA directly into the PCR product. Pipette up and down to ensure the even mixing.
2. Prepare bead hybridization mixture (requiring 16.5 μl for each detection). Mix 1 volume of 1.5× TMAC for every 0.276 volume of beads (from Subheading 3.3). Prepare enough mixture for all detections. Aliquot the bead hybridization mixture into a 96-well PCR plate (referred to as "hybridization plate"). Note that 16.5 μl of bead hybridization mixture contains ~1,500 beads per color.
3. Add 7.5 μl of PCR product (with hybridization control mix and EDTA) into each well containing the bead hybridization mixture. Pipet up and down to mix well. Do not spin the plate. Hybridize using a PCR machine with 95 °C for 5 min and 50 °C on hold. Hybridize at 50 °C for 4 h to overnight.
4. Spin down the hybridization plate at 1,600×*g* for 3 min. Remove supernatant. Add 40 μl of 1× TMAC to each well, and mix well to wash the beads. Bring down the beads by spinning again. Remove the liquid in the well (do not touch the beads; it is fine to leave ~5–10 μl liquid per well during all such washing steps).
5. Dilute SAPE stock 1:100 with 1× TMAC buffer, and add 40 μl of diluted SAPE to each well. Pipet up and down to mix well. Seal the plate and incubate at 50 °C for 10 min.
6. Spin down the beads. Wash two times with 1× TMAC. After the last wash, resuspend the beads in 50 μl per well with 1× TMAC.
7. Detect signals on beads using Luminex FlexMap 3D machine, following the manufacturer's guidelines in using the machine (*see* **Note 8**). We usually detect 100 beads per bead color.

4 Notes

1. Both MicroPlex and MagPlex beads are carboxylated, so they can covalently bind to amino-labeled DNA oligonucleotides (detection probes) in the presence of EDC. There are currently 100 colors for MicroPlex beads and up to 500 colors for MagPlex beads. Such difference will determine the grouping of bead sets (*see* also **Note 7**). In addition, MagPlex beads can

be separated from liquid using magnets rather than spinning. Both MicroPlex and MagPlex beads worked well in our hands.

2. Although a Luminex machine is recommended, it is feasible to detect the identity of the beads and signals on the beads using a regular flow cytometer. Adjustment of flow cytometer settings (such as voltages and compensation) is required for best effect.
3. We have routinely used 250 ng to 2 μg input total RNA with a high success rate of labeling. Using total RNA amount as low as 25 ng can also generate successful runs, albeit the success rate will be lower (~70 % in our hands). Increasing PCR cycles may be needed with lower input RNA amount.
4. We suggest that the layout of samples within the sample preparation plate is preplanned before the experiments. There is no need to avoid edge wells, as we have not seen problems associated with using such wells. We also recommend adding two or more wells of negative controls on each plate. These negative controls can be water-only samples that go through all the steps in labeling and detection, which can inform the background level of detection of the Luminex beads and serve as an alert in case of PCR-carryover contamination. In addition, since sample processing is by rows or columns when using multichannel pipettes, we recommend to randomize the sample location on the plate, so that samples belonging to the same biological conditions are not grouped geographically on the plate.
5. It is important that the 2× SSC washing steps are carried out stringently, so that previous rounds of washing solutions are completely removed before adding the next round of wash solution. When using a multichannel pipette, special attention should be paid to ensure that all wash solutions are removed by each of the channels.
6. This protocol uses a high concentration of 3′ adaptor and linearization of miRNAs on capture probes to avoid circularization of miRNAs in the presence of T4 RNA ligase and ATP. Alternatively, pre-adenylated 3′ adaptor can be used in combination with truncated T4 RNA ligase 2 to carry out 3′ adaptor ligation [15]. In addition to chemically synthesized pre-adenylated form of 3′ adaptor, which unfortunately is not easily available commercially at low cost, enzyme-based pre-adenylation has been reported [16].
7. Since the number of colors available for detection beads is smaller than the current number of known miRNAs, it is recommended that each PCR-labeled sample is detected in multiple detection wells, with each well having a different bead set. This concept of bead set allows using the same colored beads for different miRNAs, as long as these different miRNAs are detected in separate bead sets, effectively enabling detecting more than 500 miRNAs with this approach. It is also

recommended that the bead color to miRNA pairing is preplanned before conjugating.

8. We normally use median fluorescence intensity readings from the machine to perform data processing and normalization.

Acknowledgments

This work was supported in part by NIH grants R01CA149109 and R01GM099811 and Connecticut Stem Cell Research Fund (09SCBYALE27). We thank Eric Miska, Ezequiel Alvarez-Saavedra, Justin Lamb, David Peck, Hao Zhang, and Judy Wang for help during the process of establishing this protocol.

References

1. Bartel DP (2009) MicroRNAs: target recognition and regulatory functions. Cell 136: 215–233
2. Ambros V (2004) The functions of animal microRNAs. Nature 431:350–355
3. Garzon R, Croce CM (2008) MicroRNAs in normal and malignant hematopoiesis. Curr Opin Hematol 15:352–358
4. Farazi TA, Spitzer JI, Morozov P et al (2011) miRNAs in human cancer. J Pathol 223: 102–115
5. Lu J, Getz G, Miska EA et al (2005) MicroRNA expression profiles classify human cancers. Nature 435:834–838
6. Wang Y, Juranek S, Li H et al (2008) Structure of an argonaute silencing complex with a seed-containing guide DNA and target RNA duplex. Nature 456:921–926
7. D'Andrade PN, Fulmer-Smentek S (2012) Agilent microRNA microarray profiling system. Methods Mol Biol 822:85–102
8. Lu J, Guo S, Ebert BL et al (2008) MicroRNA-mediated control of cell fate in megakaryocyte-erythrocyte progenitors. Dev Cell 14:843–853
9. Li Z, Lu J, Sun M et al (2008) Distinct microRNA expression profiles in acute myeloid leukemia with common translocations. Proc Natl Acad Sci U S A 105:15535–15540
10. Olson P, Lu J, Zhang H et al (2009) MicroRNA dynamics in the stages of tumorigenesis correlate with hallmark capabilities of cancer. Genes Dev 23:2152–2165
11. Mi S, Lu J, Sun M et al (2007) MicroRNA expression signatures accurately discriminate acute lymphoblastic leukemia from acute myeloid leukemia. Proc Natl Acad Sci U S A 104:19971–19976
12. Adams BD, Guo S, Bai H et al (2012) An in vivo functional screen uncovers miR-150-mediated regulation of hematopoietic injury response. Cell Rep 2:1048–1060
13. Guo S, Bai H, Megyola CM et al (2012) Complex oncogene dependence in microRNA-125a-induced myeloproliferative neoplasms. Proc Natl Acad Sci U S A 109:16636–16641
14. Guo S, Lu J, Schlanger R et al (2010) MicroRNA miR-125a controls hematopoietic stem cell number. Proc Natl Acad Sci U S A 107:14229–14234
15. Hafner M, Landgraf P, Ludwig J et al (2008) Identification of microRNAs and other small regulatory RNAs using cDNA library sequencing. Methods 44:3–12
16. Vigneault F, Sismour AM, Church GM (2008) Efficient microRNA capture and bar-coding via enzymatic oligonucleotide adenylation. Nat Methods 5:777–779

Chapter 5

Using Pooled miR30-shRNA Library for Cancer Lethal and Synthetic Lethal Screens

Liam Changwoo Lee, Shaojian Gao, Qiuning Li, and Ji Luo

Abstract

Pooled shRNA library is a powerful, rapid, and cost-effective technology to carry out functional genomic screens in mammalian cells. This approach has been applied extensively to identify genetic dependencies in cancer cells that might be exploited for therapeutic purposes. In this chapter we provide a detailed protocol for using the Hannon–Elledge miR30-based library to conduct dropout screens in cancer cell lines. This protocol is readily adaptable to other pooled shRNA libraries and should facilitate the functional annotation of the human genome.

Key words RNAi, Synthetic lethality, Cancer cells, miR30-based library, Dropout screen

1 Introduction

The discovery of RNA interference (RNAi) and its application in genetic screens have revolutionized functional genomic analysis in metazoans [1–4]. Loss-of-function RNAi screens in mammalian cells using either well-by-well siRNA transfection and shRNA transduction [5] or pooled shRNA transduction [6, 7] have been particularly powerful at annotating gene functions and discovering new genetic interactions. In the area of cancer biology, RNAi-based loss-of-function screens have been widely applied for the discovery of cancer cell vulnerabilities, or "Achilles heels," that could potentially suggest new therapeutic approaches [7, 8]. RNAi screens have been directed at identifying genes whose knock down selectively kills cancer cells (cancer lethality) or subsets of cancer cells bearing particular mutations (synthetic lethality). Although much effort has been directed at screening the druggable portion of the human genome, screening the entire human genome in an unbiased fashion could result in unexpected discoveries and new insights, especially with genes that are previously poorly characterized. A major advantage of using pooled shRNA libraries is its versatility and low cost, which enables the rapid screening of many

Narendra Wajapeyee (ed.), *Cancer Genomics and Proteomics: Methods and Protocols*, Methods in Molecular Biology, vol. 1176, DOI 10.1007/978-1-4939-0992-6_5, © Springer Science+Business Media New York 2014

cancer cell lines [9]. This provides a cost-effective and rapid way to functionalize the cancer genome for the identification of new drug targets.

Recently we have reviewed some of the general considerations for using pooled shRNA library in screens [10]. In this chapter we provide a detailed protocol for using the Hannon–Elledge miR30-based shRNA library [11] in the MSCV-PM retroviral vector for pooled shRNA screens. This library has been applied successfully in dropout screens to identify cancer lethal [7] and synthetic lethal [12] genes. It has also been used in enrichment screens to identify tumor suppressors [13, 14] and antiproliferative genes [15]. The shRNAs in this library are commercially available through Open Biosystems (Thermo Scientific) with several delivery vectors (retroviral, lentiviral, and tet inducible). Methods described in this protocol require only common laboratory equipment and techniques and thus can be carried out by most academic laboratories without the need for high-throughput robotics. Furthermore, this protocol can be readily adapted for any pooled shRNA retroviral and lentiviral library.

2 Materials

2.1 Plasmids, Cell Lines, Chemicals, and Other Reagents

1. shRNA library plasmid stock.
2. MSCV-PM-FF2 shRNA (sense strand cccgcctgaagtctctgattaa).
3. MSCV-PM-TSC2 shRNA#2 (sense strand agcctgcccttccggaaggatt).
4. MSCV-PM-TSC2 shRNA#3 (sense strand acctgtcagtgaaataaataaa).
5. Library-grade competent bacteria (>1×10^{10} transformants/μg plasmid DNA).
6. 50 μg/mL LB-agarose plate with ampicillin.
7. 50 μg/mL LB-broth with ampicillin.
8. Glycerol (cell culture grade).
9. Plasmid Plus Maxi kit.
10. 293T cells.
11. U2OS cells.
12. DMEM media.
13. McCoy's 5A media.
14. Opti-MEM media.
15. Fetal bovine serum (FBS).
16. Penicillin/streptomycin supplement.
17. TransIT-293 transfection reagent.
18. 8 mg/mL Polybrene, cell culture grade.
19. 1 mg/mL Puromycin, cell culture grade.

20. TSC2 antibody.
21. Hot-start Taq polymerase.
22. DNA gel purification kit.

2.2 Equipment

1. Electroporator (for electroporating competent bacteria).
2. 37 °C Incubator with shaker (for growing liquid bacterial culture).
3. Spectrophotometer (for monitoring OD of liquid bacterial culture).
4. Cell culture incubator (5 % CO_2, 95 % humidity, 37 °C) and biosafety hood (for cell culture works).
5. Beckman centrifuge with an SX4750 rotor/adaptor set or equivalent.
6. Thermocycler.
7. DNA agarose gel electrophoresis apparatus.

2.3 Oligonucleotides

Oligo name	Sequence
JH353F	tagtgaagccacagatgta
BC1R	cctcccctacccggtaga
P7-loop	agcagaagacggcatacgatagtgaagccacagatgta
P5-mir3A	aatgatacggcgaccaccgagctcctaaagtagccccttgaattccgaggcagtaggca
mir30-seq2	gctcctaaagtagccccttgaattccgaggca

3 Methods

3.1 Library Propagation

The initial design and generation of a large shRNA library consisting of tens of thousands of shRNA clones [5, 11] are beyond the scope of this review and the technical capacity of most individual labs and therefore are not be discussed here. However, once generated, pooled shRNA library plasmids can be maintained and amplified easily. A major quality control during library amplification is maintaining the representation of individual shRNAs in the plasmid pool.

1. Transform 100 ng of library plasmids into high-competency *E. coli* by electroporation (for example, with the ElectorMax DH10B library-grade competent bacteria according to the manufacturer's instruction).
2. Take 1 % of the transformed bacteria and carry out tenfold serial dilution plating to estimate transformation efficiency and total

transformant count. Total number of transformants should be ≥100,000-fold of pool complexity. For example, for a 3,000 shRNA plasmid pool, aim for $\geq 3 \times 10^8$ transformants.

3. Plate the entire transformed bacterial culture in one 15 cm LB-agar plate containing 50 μg/mL ampicillin. We use glass beads to spread the culture evenly on the plate. Incubate overnight at 37 °C.
4. The next day there should be an even lawn of bacteria on the LB-agar plate. Pre-warm a flask of 100 mL LB broth containing 50 μg/mL ampicillin to 37 °C. Add 5 mL of LB broth to the plate using a cell scraper to gently scrape off bacteria, and transfer to the flask. Repeat the collection process twice more to collect all bacteria from the plate.
5. Grow the bacteria culture in a 37 °C incubator with shaking (250 rpm) for ~4 h. Monitor O.D. of the culture periodically until it reaches ~0.6.
6. Freeze down 1 mL aliquots in 10 % glycerol for future library propagation.
7. Maxi-prep the entire culture.
8. Optional: The shRNA composition of the amplified plasmid pool can be verified by deep sequencing (*see* Subheading 3.7 below).

3.2 Library Production

We package the MSCV-PM-shRNA library with a plasmid co-transfection method in 293T cells using TransIT-293 transfection reagent (Mirus Bio). To maintain complexity of the pooled shRNA library, it is vital that both the transfection efficiency of 293T cells is high and the viral titer in the resulting supernatant is high. The yield of the shRNA library virus significantly depends on the purity of the plasmids and the conditions of the 293T cells. Therefore, maxi-prep-grade plasmids are strongly recommended and 293T cells must be maintained in log phase.

1. From a log-phase stock of 293T cells (grown in DMEM + 10 % FBS + P/S), plate cells in antibiotic-free media (DMEM + 10 % FBS) in 10 cm cell culture plates at 5.5×10^6 cells/plate.
2. 293T cells should reach 75–80 % confluency in the afternoon of the second day. Good cell density will prevent cells from lifting off the plate after transfection.
3. Prepare a mixture of the shRNA library plasmid together with vsv-G and Gag/Pol retroviral packaging plasmids at 5:1:1 ratio. For each 10 cm plate use 5 μg of shRNA library plasmid and 1 μg each for vsv-G and Gag/Pol plasmids.
4. To generate control shRNAs, prepare mixture of vsv-G and Gag/Pol plasmid with the following plasmids as in **step 3**: MSCV-GFP (titering control), MSCV-PM-FF2 (negative control

shRNA targeting firefly luciferase), and MSCV-PM-TSC2 shRNA#2 and shRNA#3 (knockdown control).

5. For each 10 cm plate, mix 21 μl of TransIT-293 reagent in 1 mL serum-free Opti-MEM media and gently vortex to mix. Incubate at room temperature for 15 min.
6. Add the plasmid mixture from **step 3** above to TransIT-293/opti-MEM, and gently vortex to mix. Incubate at room temperature for 30 min.
7. Remove media from the plate of 293T cells to leave 5 mL behind. Add the transfection mixture dropwise to the cells, and gently rock to mix.
8. Incubate cells overnight in cell culture incubator.
9. In the morning of the third day replace the media with 6 mL of DMEM + 10 % FBS to remove residual transfection reagent.
10. Incubate cells overnight in cell culture incubator.
11. In the morning of the fourth day, collect the 6 mL of media (which contains viruses) and store at 4 °C in a sterile 50 mL conical tube. Replace with 6 mL of fresh DMEM + 10 % FBS media. Care should be taken to minimize dislodging of the cells. Collect the media again 8–12 h later, combine with previous collection, and gently invert to mix.
12. Centrifuge the viral supernatant at 137 × *g* (800 rpm with a Beckman SX4750 rotor) for 5 min to pellet any 293T cells.
13. Aliquot the clarified viral supernatant in 5 mL aliquots and store at −80 °C. Viral supernatant should be good for 1 year. Also set aside several 20 μl aliquots and store at −80 °C for titering. There is no need to filter the viral supernatant as complete freezing will kill any contaminating 293T cells.

3.3 Library Titering

We routinely use U2OS cell colony assay to titer the library virus because U2OS cells have high clonogenic efficiency. Other highly clonogenic cell lines such as HeLa may also be used. Titer determined from these reference cell lines must be adjusted for the cell lines used in the screen (*see* Subheading 3.4 below). We typically obtain library titer of $1–5 \times 10^6$ cfu/mL in U2OS cells.

1. From a log-phase stock of U2OS cells grown in complete media (McCoy's 5A media + 10 % FBS + P/S), plate cells in 10 cm plate at 1.5×10^6 cells/plate and incubate overnight in cell culture incubator.
2. On the second day, thaw out a 20 μl aliquot of library (or control shRNA) viral supernatant and mix 1 μl of virus with 6 mL of complete media containing 4 μg/mL polybrene. Replace the media on U2OS cells with the virus–polybrene–media mixture. Incubate overnight.

3. On the third day, split each plate of cells 1:20 into duplicate 10 cm plates.
4. Incubate cells overnight in cell culture incubator.
5. On the fourth day, replace media with complete media containing puromycin (1 μg/mL).
6. Incubate cells for 7–10 days, and refresh selection media every 3 days until colonies become visible.
7. Wash cells with DPBS once, and stain cells with Coomassie blue. Count the number of colonies. Viral titer is *colony count* × 10,000 cfu/mL.

3.4 Cell Line Verification

Cell lines that are to be used for the screen should be verified both for their infection efficiency and their RNAi response. For a given MSCV pseudotype different cell lines will have different infection efficiency. As the library is routine titered on U2OS cells, the "relative titer" of the library for a specific cell line must be determined empirically with an MSCV-GFP control virus of the same pseudotype. Although miR30-based shRNAs work well in most human and mouse cell lines, each cell line must be verified for their shRNA knockdown efficiency. We routinely used two shRNAs against the tumor suppressor TSC2 for verifying human cell lines. One shRNA provides strong knockdown, whereas the other shRNA provides weak knockdown at single-copy integration; thus they provide a semiquantitative assessment of a cell line's response to miR30-shRNA.

1. Generate the following control viruses, and titer them in U2OS cells as in Subheadings 3.2 and 3.3: MSCV-GFP, MSCV-PM-FF2, and MSCV-PM-TSC2 shRNA#2 and shRNA#3.
2. To measure the relative titer of a cell line of interest, plate log-phase cells at a density of ~100,000/well in 6-well plates (initial cell number should be adjusted empirically such that cells do not reach confluency within 48 h).
3. Immediately before infection, count one well to determine the actual cell number at the time of infection. Infect cells with the MSCV-GFP virus at various multiplicity of infection (MOI): 0.1, 0.2, 0.5, 1, 2, and 4 by mixing the appropriate amount of MSCV-GFP viral supernatant with cell media in a total volume of 1.2 mL. Polybrene at 4 μg/mL works for most cell lines but should be determined empirically to minimize toxicity.
4. Optional: Spin infection can be attempted to improve transduction efficiency for cell lines that are hard to infect. For spin infection, centrifuge 6-well plates at 930 × *g* (2,000 rpm with a Beckman SX4750 rotor) for 30 min.
5. Change media the next morning. For cell lines that do not tolerate prolonged serum exposure (such as primary epithelial

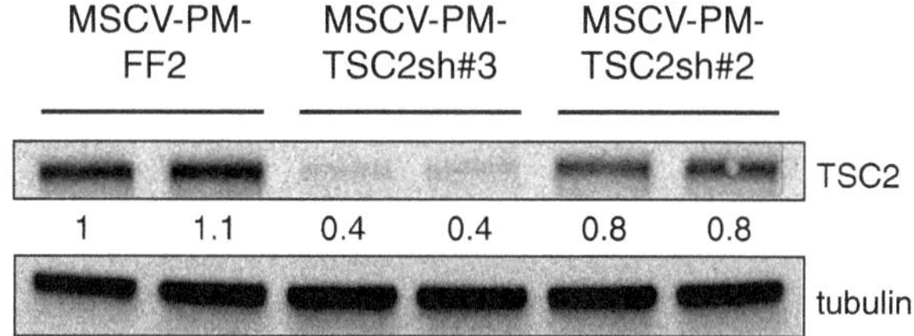

Fig. 1 Control TSC2 shRNAs for verifying that a cell line is sensitive to miR30-shRNA-mediated gene knockdown. These two shRNAs were chosen to give weak (20–30 %) and strong (>60 %) knockdown of TSC2 at single-copy integration in the reference U2OS cell line

cells), the infection time may be shorted to 4 h with a media change afterwards.

6. 48 h after infection, quantify the % of GFP+ cells under each MOI condition. This can be done with either a standard inverted fluorescence microscope or a FACS analyzer. With increasing MOI there should be a linear range of increasing % of GFP+ cells before a plateau where most cells became GFP+. Use the linear range of the data to estimate the relative titer of the viral pseudotype. For example, if at an MOI of 0.2 there are 10 % of GFP+ cells and at MOI of 0.5 there are 25 % of GFP+ cells, then for this cell line its *relative titer* = 0.5× *U2OS titer*.
7. To test miR30-shRNA knockdown efficiency in a specific cell line, plate log-phase cells in 6-well plates and infect with the negative control shRNA MSCV-PM-FF2 and the two TSC2 shRNAs at MOI of 1 (follow **steps 2–5** above) based on the relative titer.
8. Two days after infection, carry out puromycin selection for 3–4 days to remove uninfected cells (puromycin concentration must be determined empirically).
9. Once puromycin selection is complete, collect whole-cell lysates and carry out Western blot for TSC2 proteins. TSC2 shRNA#2 typically gives 20–30 % knockdown, and TSC2 shRNA#3 typically gives 60–80 % knockdown (Fig. 1).

3.5 shRNA Library Dropout Screen

An shRNA dropout screen works well if cells behave homogeneously and there is sufficient representation of the library in the population. For a more detailed discussion *see* ref. 10. We recommend a representation of ~1,000 (*see* **Note 1**). Care must be taken that the representation of the library is maintained throughout the screen. The number of population doubling (PD) during the screen should also be determined beforehand (*see* **Note 2**). As with all high-throughput screens, multiple biological replicates are necessary.

The protocol below was based on a library of 3,000 shRNAs with a representation of 1,000 (i.e., an average of 1,000

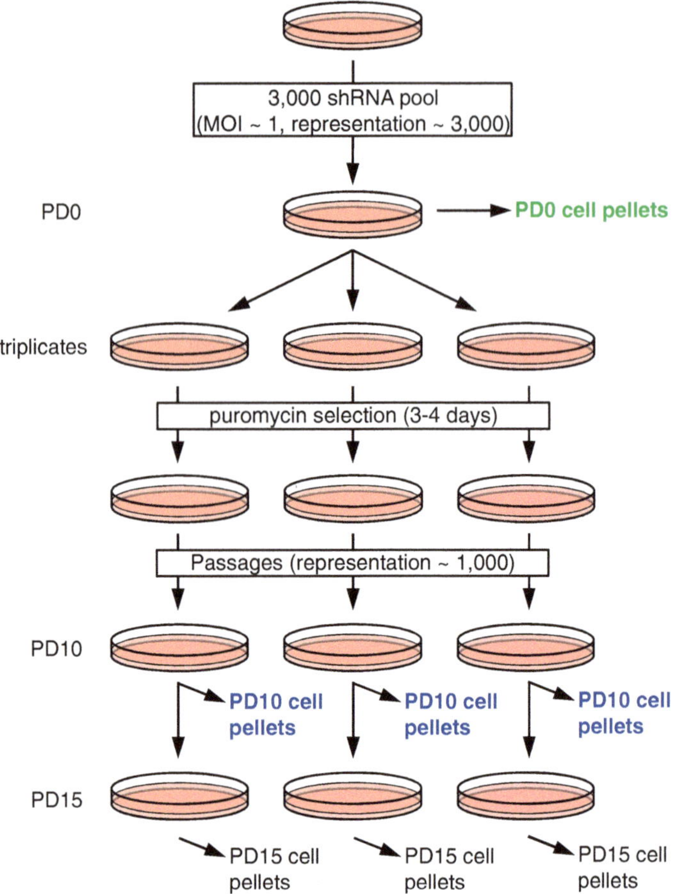

Fig. 2 Schematic outline of a pooled shRNA screen using a library of 3,000 shRNAs in triplicates

independent integrations per shRNA). For larger libraries, scale virus and cell numbers proportionally. An outline of the protocol is shown in Fig. 2. As the relative depletion/enrichment of shRNAs in the pool depends on the number of PDs (*see* **Note 2**), for inter-cell line comparisons (isogenic cell lines or panel of cancer and normal cell lines), it is important that the assay is done with the same number of PDs for all cell lines.

1. Plate log -phase cells in 15 cm plates at the appropriate density such that they do not reach confluency within 48 h.
2. Immediately before infection, trypsinize one plate and count cell number at the time of infection.
3. Infect the shRNA library into cells at an MOI of 1. Transduce a total of 9×10^6 viral particles into cells (average representation 3,000/shRNA). Viral transduction can be spread out over multiple 15 cm plates.

4. Two days after transduction, trypsinize and pool all cells. Plate triplicates with 3×10^6 cells per replica (average representation 1,000/shRNA). Cells may be spread out over multiple 15 cm plates at 10–20 % confluency to enable 2–3 PDs in log phase between passages. Freeze down the rest of the cells in two or more cell pellets with 3×10^6 cells per pellet. These are the PD0 samples for genomic DNA extraction.
5. Propagate each replica independently. Select cells with puromycin for 3–4 days to remove any uninfected cells. For each replica, when cells reach 70–80 % confluency, trypsinize cells and record cumulative PD. Passage 3×10^6 cells for propagation.
6. At PD = 10, after passaging, freeze down the leftover cells from each replica as PD10 samples for genomic DNA extraction.
7. At PD = 15 (or a pre-determined final PD), freeze down all cells from each replica as PD15 samples for genomic DNA extraction.

3.6 Library Recovery and Deep Sequencing

To identify shRNAs that negatively (and positively) affect cell viability and proliferation rate, shRNA sequences are recovered from genomic DNAs by PCR amplification and the compositions of the library at different time points are determined by deep sequencing. PCR strategies can be developed to recover either the entire hairpin sequences or only half of the hairpin sequences. We initially deconvolved the library using custom barcode microarrays [7]. As deep-sequencing platforms become cheaper and more accessible, direct sequencing of the shRNAs is now the preferred method for library deconvolution. The main advantages of deep sequencing over barcode microarray are that it does not suffer from the problem of ratio compression for highly depleted/enriched shRNAs and that it eliminates the issue of probe specificity associated with microarrays. To generate an accurate count of shRNAs in the library, it is necessary to sequence the library at sufficiently high coverage. We recommend a sequencing coverage of ≥500-fold (i.e., $\geq 1.5 \times 10^6$ reads per sample for a 3,000 shRNA library). To prepare shRNA sequences for Illumina sequencing, we used a two-step PCR (Fig. 3). The first step is to recover shRNA half-hairpins from genomic DNA. We chose to amplify only half of the shRNA hairpin to avoid intramolecular annealing of hairpin sequences that could adversely affect the sequencing process. The second step is to add Illumina adaptors to the PCR amplicon.

1. Prepare genomic DNA from PD0 and PD15 cell pellets (for example, using Qiagen Blood and Cell Culture DNA Maxi Kit). Use at least 3×10^6 cells per prep (i.e., average representation 1,000/shRNA). Avoid excessive shredding of genomic DNA, but the DNA preparation should also not be viscous when diluted to 1 mg/mL.

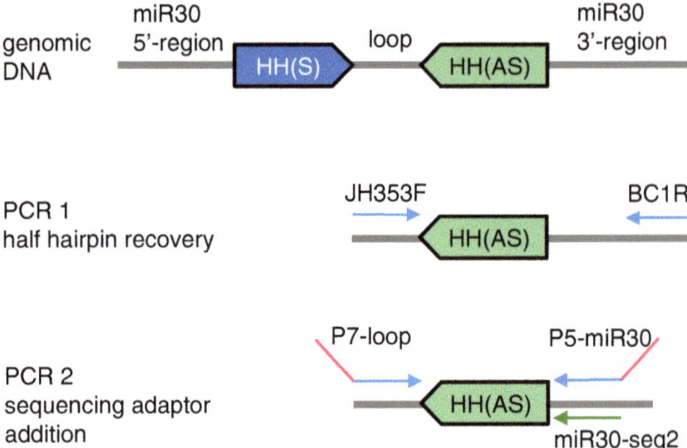

Fig. 3 Schematic outline of the two-step PCR protocol that recovers shRNA half-hairpin from genomic DNA and adapts Illumina sequencing linkers to the amplicon. HH(S), half-hairpin sense; HH(AS), half-hairpin antisense; miR30-seq2, custom Illumina sequencing primer

2. For each genomic DNA sample, set up the following PCR reaction to recover shRNA half-hairpins from genomic DNA. Split the master reaction mixture into 100 μl aliquots.

Component	Volume	Note
10× PCR Buffer (Takara)	80 μl	
2.5 mM 10× dNTP	64 μl	Final conc. 0.20 mM each
100 μM Primer JH353F	4 μl	Final conc. 0.5 μM, anneals to miR30 hairpin loop
100 μM Primer BC1R	4 μl	Final conc. 0.5 μM anneals to miR30 3′ constant region
DMSO	16 μl	Final conc. 2 %
5 U/μl Hot-start Taq	8 μl	It is important to use a hot-start Taq to ensure specific and efficient amplification of shRNA half-hairpins
Genomic DNA	Up to 60 μg	Assuming that each human diploid genome is ~6.5 pg, use 6.5 ng of DNA per shRNA. Thus for a 3,000 shRNA library use 19.5 μg of DNA
Dd H_2O	To 800 μl	
Total	800 μl	

PCR program:

Step	Temperature	Time
1	95 °C	5 min
2	94 °C	35 s
3	52 °C	35 s
4	72 °C	1 min
5		Go to step 2 for 34 more cycles
6	72 °C	10 min

3. Precipitate DNA from the PCR reaction (equal volume isopropanol, 150 mM sodium acetate pH 5.2) at −20 °C for 1 h. Wash the DNA pellet with 75 % ethanol, and resuspend 50 μl of 1× TE buffer.
4. Electrophorese the PCR product in a 1.5 % agarose TAE gel, and purify the ~320 bp PCR amplicon (for example, using Qiagen Gel Extraction Kit).
5. Set up the following PCR reaction to add Illumina adapter primers.

Component	Volume	Note
10× PCR Buffer (Takara)	10 μl	
10× dNTP (2.5 mM each)	10 μl	Final conc. 0.25 mM each
100 μM Primer P-7 loop	2 μl	Final conc. 0.2 μM
100 μM Primer P5-mir3A	2 μl	Final conc. 0.2 μM
5 U/μl Hot-start Taq	2 μl	
Input DNA	500 ng	
ddH_2O	To 100 μl	
Total	100 μl	

PCR program:

Step	Temperature	Time
1	95 °C	5 min
2	95 °C	15 s
3	50 °C	30 s
4	72 °C	30 s
5		Go to step 2 for 4 more cycles
6	95 °C	15 s

(continued)

Step	Temperature	Time
7	56 °C	30 s
8	72 °C	30 s
9		Go to step 6 for 9 more cycles
10	72 °C	

6. Electrophorese the PCR product in a 3 % Nusieve agarose TAE gel, and purify the 114 bp PCR product (for example, using Qiagen Gel Extraction Kit).
7. Sequence the shRNA half-hairpins using the mir30-seq2 primer. We typically use the Illumina platform with 36 nt read length.

3.7 Bioinformatics

We map all qualified reads from the Illumina sequencing reaction to an shRNA sequence database and count the number of reads for each shRNA. A fraction of shRNA reads may have mismatches due to PCR/sequencing errors. We have compared the results using reads with only perfect match vs. reads with 0 or 1 mismatch and the results are similar.

Next the PD15 samples are compared to PD0 samples, and a normalized PD15-to-PD0 log2 ratio for each shRNA is calculated. Due to the nature of shRNA competition in the pool, it is possible that shRNAs with neutral effect will decrease or increase in their representation over time depending on how many shRNAs are enriching or dropping out, respectively. We assume that the majority of shRNAs have neutral behavior in the assay and normalize the dataset such that the mean PD15:PD0 log2 ratio is 0. We apply statistical analysis of microarray (SAM) [7] test to identify shRNAs that have dropped out or enriched in the screen, with a typical cutoff ≥2-fold and an FDR of 10 %. Alternatively, hits can be identified using *z* scores [16]. The Broad Institute has developed an algorithm, RIGER, that is more tailored to analyzing deep shRNA libraries [8].

4 Notes

1. *Considerations on representation and MOI*:
For dropout screens, we typically use an MOI of ~1 and a representation of 1,000. A sufficiently high number of independent retroviral integrations are necessary to ensure (1) that shRNAs that are presented at lower copy number in the library are sufficiently represented in the screen and (2) averaging out any integration positional effect for individual shRNAs. We have attempted dropout screens at representation of ~500.

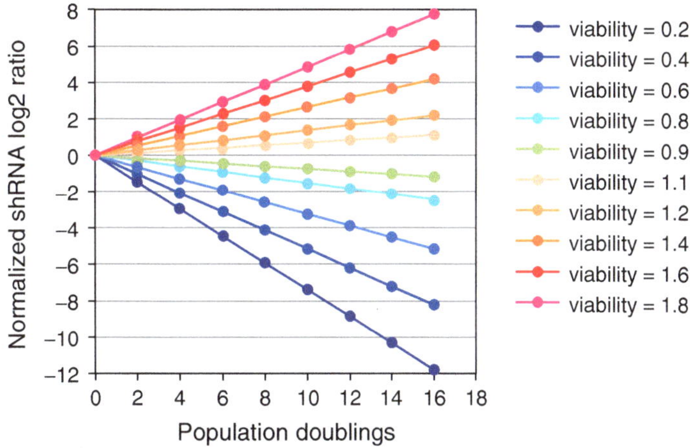

Fig. 4 Simulation of shRNA dropout and enrichment with a pool of 100 shRNAs that contains 90 "neutral" shRNAs (per cell doubling viability = 1) and 10 "hit" shRNAs that either negatively (per cell doubling viability < 1) or positively (per cell doubling viability > 1) affect cell proliferation. The population was simulated with resampling every 2 PDs at a representation of 1,000. Changes in normalized log2 ratio of each hit shRNA at a given PD vs. PD0 are plotted

We do not recommend representations below 500 for dropout screens.

2. *Considerations on optimal population doubling for the screen*: One important factor for dropout screen is to consider the optimal number of PDs for the screen. Those shRNAs with strong negative effects on cell viability and proliferation will drop out within a few PDs, whereas those with weak negative effects on cell viability will drop out over longer periods of time. As current shRNA libraries, including the miR30-shRNA library discussed here, are not validated for knockdown, an shRNA with a weak effect on cell viability could therefore result from modest target knockdown. Thus the assay duration needs to be sufficiently extensive to detect the dropout of weaker shRNAs. Most shRNA libraries are likely to also contain shRNAs that positively influence cell proliferation (such as those targeting tumor suppressor and antiproliferative genes); these shRNAs will enrich over time. Since each cell passage represents a resampling of the library, too many PDs could result in the over-enrichment of growth-promoting shRNAs and a reduction of both negative and neutral shRNAs in the pool. This could adversely affect data analysis. Figure 4 illustrates a simulation of the behavior of 5 shRNAs that negatively affect cell proliferation and 5 shRNAs that positively affect cell proliferation in a pool with 90 neutral shRNAs. We typically carry the screen to ~15 PDs and collect an intermediate PD10 sample. This allows us to look at two time points if necessary to identify both strong and weak hits.

Acknowledgement

This work was supported by a National Cancer Institute Center for Cancer Research Intramural Grant to J.L.

References

1. Mello CC, Conte D (2004) Revealing the world of RNA interference. Nature 431:338–342
2. Chang K, Elledge SJ, Hannon GJ (2006) Lessons from Nature: microRNA-based shRNA libraries. Nat Methods 3:707–714
3. Root DE, Hacohen N, Hahn WC, Lander ES, Sabatini DM (2006) Genome-scale loss-of-function screening with a lentiviral RNAi library. Nat Methods 3:715–719
4. Bernards R, Brummelkamp TR, Beijersbergen RL (2006) shRNA libraries and their use in cancer genetics. Nat Methods 3:701–706
5. Moffat J et al (2006) A lentiviral RNAi library for human and mouse genes applied to an arrayed viral high-content screen. Cell 124:1283–1298
6. Ngo VN et al (2006) A loss-of-function RNA interference screen for molecular targets in cancer. Nature 441:106–110
7. Schlabach MR et al (2008) Cancer proliferation gene discovery through functional genomics. Science 319:620–624
8. Luo B et al (2008) Highly parallel identification of essential genes in cancer cells. Proc Natl Acad Sci U S A 105(51):20380–20385
9. Cheung HW et al (2011) Systematic investigation of genetic vulnerabilities across cancer cell lines reveals lineage-specific dependencies in ovarian cancer. Proc Natl Acad Sci U S A 108:12372–12377
10. Hu G, Luo J (2012) A primer on using pooled shRNA libraries for functional genomic screens. Acta Biochim Biophys Sin (Shanghai) 44:103–112
11. Silva JM et al (2005) Second-generation shRNA libraries covering the mouse and human genomes. Nat Genet 37:1281–1288
12. Luo J et al (2009) A genome-wide RNAi screen identifies multiple synthetic lethal interactions with the Ras oncogene. Cell 137: 835–848
13. Westbrook TF et al (2005) A genetic screen for candidate tumor suppressors identifies REST. Cell 121:837–848
14. Zender L et al (2008) An oncogenomics-based in vivo RNAi screen identifies tumor suppressors in liver cancer. Cell 135:852–864
15. Solimini NL et al (2012) Recurrent hemizygous deletions in cancers may optimize proliferative potential. Science 337:104–109
16. Birmingham A et al (2009) Statistical methods for analysis of high-throughput RNA interference screens. Nat Methods 6:569–575

Chapter 6

A Diphtheria Toxin Negative Selection in RNA Interference Screening

Zhi Sheng, Susan F. Murphy, Sujuan Guo, and Michael R. Green

Abstract

RNA interference (RNAi) screening is a powerful technique for understanding the molecular biology of cancer and searching drug targets. Genes and their upstream activators that are essential for the survival of cancer cells often dictate cancer formation/progression. Hence, they are preferable therapeutic targets. Identifying these genes using RNAi is, however, problematic because knocking them down leads to cell death. Here we describe a diphtheria toxin (DT) negative selection method to circumvent the problem of cell death in RNAi screening. DT fails to kill mouse cells due to the lack of functional DT receptor (DTR). Thus, we first prepare a construct encoding a human functional DTR driven by the promoter of mouse *Atf5*, a gene essential for the survival of malignant glioma. Then a DT-sensitive mouse malignant glioma cell line is established by over-expressing this DTR. Finally, an RNAi screen is performed in this cell line and genes that activate *Atf5* expression are identified. The negative selection approach described here allows RNAi screening to be used for identifying genes controlling cell survival in cancers or perhaps other human diseases with potential in therapeutic intervention.

Key words Diphtheria toxin, Diphtheria toxin receptor, Negative selection, RNA interference screen, Activating transcription factor 5, Malignant glioma

1 Introduction

Short-hairpin RNA (shRNA)-based RNA interference (RNAi) screening, first developed in 2004 [1], utilizes a library of DNA constructs encoding shRNA that are capable of knocking down their target genes. Hence, an RNAi screen identifies candidates directly based upon the function and/or activity of genes in a genome [2]. An important step in successfully carrying out an RNAi screen is isolating the cells containing shRNAs of interest. The most frequently used approach for this purpose is positive selection, which is ideal only for identifying genes that silence their targets or induce cell death because knocking down these genes activates their targets or retains cell viability permitting a positive selection. For example, in a KRas-transformed NIH3T3

Narendra Wajapeyee (ed.), *Cancer Genomics and Proteomics: Methods and Protocols*, Methods in Molecular Biology, vol. 1176, DOI 10.1007/978-1-4939-0992-6_6, © Springer Science+Business Media New York 2014

cells, the Fas gene is silenced by KRas. An RNAi screen identified shRNAs that reverse KRas-mediated Fas gene silencing by positively selecting Fas-expressing cells [3]. Oncogenes induce senescence (a type of cell death) during tumor transformation [4]. Wajapeyee et al. performed an RNAi screen, in which they isolated human primary foreskin fibroblast (PFF) cells that survived from the senescence induced by the oncogene BRAFV600E. From these PFF cells, they identified IGFBP7 as a potent senescence inducer, which has profound implications in cancer therapeutic intervention [5].

Nevertheless, it is difficult to utilize positive selection to screen genes required for cell survival due to the induction of cell death. For instance, activating transcription factor 5 (ATF5) is overexpressed in and controls the survival and proliferation of malignant glioma cells [6]. Knocking down ATF5 induces apoptosis [7]. Thus, alternative strategies in RNAi screening need to be taken to study ATF5-mediated survival pathways in malignant glioma. To address this technical difficulty, we describe here a negative selection approach in which we take advantage of the selective cytotoxicity of diphtheria toxin (DT) to human but not mouse cells. DT, a polypeptide secreted by *Corynebacterium diphtheriae*, kills human cells as they express DT receptor (DTR) (also known as heparin-binding EGF-like growth factor). Unlike human cells, mouse cells are resistant to DT due to the lack of functional DTR [8]. Thus, exogenously expressing a functional DTR converts DT-resistant mouse cells into DT-sensitive cells and silencing this DTR reverses it; such a reporter system therefore endorses an isolation of DTR-lacking/DT-resistant mouse cells.

To build and test this reporter system, a promoter of mouse *Atf5* gene is cloned in front of a human DTR gene fused with EGFP. The resulting construct p*Atf5p*-DTR-EGFP is used to establish a DT-sensitive mouse malignant glioma cell line. An RNAi screen is performed using DT negative selection, and *Atf5* transcriptional activators are identified. This negative selection reporter is a feasible and complementary tool in RNAi screening and broadens the application of this technique in the research of cancers and perhaps other human diseases.

2 Materials

2.1 Cloning

1. MCD11C-DTR-EGFP: This plasmid was kindly provided by Dr. Richard Lang at the Cincinnati Children's Hospital [9].
2. PCR primers: Reconstitute the oligos to 20 pmol/μl in deionized, sterile water. Store the reconstituted oligos at −20 °C. DTR-NT-RI and EGFP-N1-CTXba1 are the primers for

amplifying DTR-EGFP fusion gene. *Atf5p*-4117 bp and *Atf5p*-Rev are the primers for amplifying 4117 bp of mouse *Atf5* promoter. Oligonucleotide sequences are:

DTR-NT-RI: 5′-ACCATGAAGCTGCTGCCGTCG-3′.

EGFP-N1-CT-XbaI: 5′-TTACTTGTACAGCTCGTCCATGCC-3′.

Atf5p-4117 bp: 5′-GCCTGTGCCCTTTCACCACTAAC-3′.

Atf5p-Rev: 5′-GGGAGCGTTAAGGGAGACTGTGCTGAG-3′.

3. pMyc-His B promoter-less vector: Digest 5 μg of the plasmid pEF4-Myc-His B (Invitrogen) with 50 U of MluI and Acc65I in buffer D (Promega) at 37 °C for 3 h. Then add 2.5 U of calf intestinal alkaline phosphatase (CIP), and incubate at 37 °C for 1 h. Purify the digested and dephosphorylated pMyc-His B vector using QIAGEN gel purification kit. Mix the vector with in vitro-synthesized and annealed oligos containing multiple cloning sites (MluI, BstZI, SacI, XhoI, EcoRV, NotI, and Acc65I) (*see* **Note 1**) at a molar ratio of 1:10, 5 μl of 2× ligase buffer, and 1 U of T4 DNA ligase (Promega). Incubate at 15 °C overnight. Transform ligated DNAs into competent DH5α cells. Prepare DNA using Wizard® Plus SV Minipreps DNA Purification System (Promega), and identify positive clones using restriction enzyme digestion. The resulting plasmid is designated as pMyc-His B.
4. pGL4.14-*Atf5p*-4 kb: 4117 bp of mouse *Atf5* promoter is amplified from a BAC clone using PCR and cloned into the pGL4.14 luciferase reporter vector (Promega). This promoter is capable of driving the expression of a luciferase reporter [10].

2.2 Cells

1. GL261: GL261 is a mouse malignant glioma cell line kindly provided by Dr. Yancey Gillespie at the University of Alabama. Cells are maintained in Dulbecco's modified Eagle's medium (DMEM)/F12 (1:1) media supplemented with 10 % of fetal bovine serum (FBS), 100 U/ml of penicillin, and 100 μg/ml of streptomycin.
2. U251: U251 is a human glioblastoma cell line purchased from ATCC. Cells are maintained in DMEM supplemented with 10 % of FBS, 100 U/ml of penicillin, and 100 μg/ml of streptomycin.
3. GL261/pCMV-3xFLAG-ATF5 cells (*see* **Note 2**): Transfect GL261 cells with the plasmid pCMV-3xFLAG-ATF5 linearized by ScaI. Select GL261/pCMV-3xFLAG-ATF5 cells with G418 selection media (0.5 mg/ml) (Subheading 2.4, **item 1**), and maintain the established cell line in G418-maintaining media (0.25 mg/ml) (Subheading 2.4, **item 2**).

2.3 DT Cytotoxicity

1. DT-dissolving buffer: Sodium phosphate buffer (10 mM, pH 7.4), 5 % lactose. Filter through a 0.45 μm Corning filter. Store at room temperature (RT).
2. DT stock solution (20 μM): Dissolve 1 mg of DT (Sigma) (MW ~ 63,000) in 0.794 ml of DT-dissolving buffer to make up 20 μM stock solution. Make 20 μl aliquots, and store at −80 °C (*see* **Note 3**). Avoid frequent thaw and freeze.
3. Crystal violet solution (0.5 %): Dissolve 0.5 g of crystal violet in 25 ml of methanol and 75 ml of deionized water. Filter through 3 M Whatman filter paper. Store at RT.

2.4 Cell Line Generation

1. G418 (0.25 mg/ml)-maintaining media: Add G418 (10 mg/ml) (Invitrogen) 1:40 into DMEM/F12 (1:1) with 10 % of FBS, 100 U/ml of penicillin, and 100 μg/ml of streptomycin (*see* **Note 4**).
2. Zeocin (50 mg/ml) stock solution: Dissolve 1 g of zeocin in 20 ml of deionized water. Make 1 ml aliquots, and store at −80 °C.
3. Zeocin (50 μg/ml) selection media (make fresh): Add zeocin (50 mg/ml) stock solution (1:1,000) to DMEM/F12 (1:1) supplemented with 10 % of FBS, 100 U/ml of penicillin, and 100 μg/ml of streptomycin.
4. G418 (0.25 mg/ml)- and zeocin (20 μg/ml)-maintaining media (make fresh): Add G418 (10 mg/ml) (Invitrogen) 1:40 and zeocin (50 mg/ml) stock solution (1:2,500) to DMEM/F12 (1:1) supplemented with 10 % of FBS, 100 U/ml of penicillin, and 100 μg/ml of streptomycin (*see* **Note 4**).
5. DT (20 nM) working solution: Dilute DT (20 μM) 1:1,000 with DT-dissolving buffer (Subheading 2.3, **item 1**). Aliquot, and store at −80 °C (*see* **Note 3**).
6. DT (100 pM) selection media (make fresh): Add DT (20 nM) working solution (1:200) to DMEM/F12 (1:1) supplemented with 10 % of FBS, 100 U/ml of penicillin, and 100 μg/ml of streptomycin.

2.5 RNAi Screening

1. shRNA library virus preparation: Seed 2×10^6 HEK293T cells in DMEM media supplemented with 10 % of FBS, 100 U/ml of penicillin, and 100 μg/ml of streptomycin in a 6-well plate coated with poly-D-lysine (Millipore) (*see* **Note 5**). Transfect cells with the plasmids containing a library of shRNAs targeting mouse genes (*see* **Note 6**) together with psPAX2 (Packaging) (Addgene) and pMD2.g (VSVG, envelope) (Addgene) in a ratio of 1:1:0.5 (*see* **Note 7**) using Effectene (QIAGEN). 24 h later, feed cells with 2.5 ml of fresh media. After another 24-h incubation, filter the media (viral supernatant) with a 0.45 μm Corning filter. Aliquot, and store at −80 °C (*see* **Note 8**).

2. Polybrene (10 mg/ml) stock solution: Dissolve 1 g of polybrene (hexadimethrine bromide, Sigma) in 100 ml of deionized, sterile water. Aliquot, and store at −80 °C.
3. Puromycin (10 mg/ml) stock solution: Dissolve 10 mg of puromycin dihydrochloride (Sigma) in 1 ml of sterile water. Aliquot, and store at −20 °C.
4. Puromycin (1 μg/ml) selection media: Add puromycin stock solution 1:10,000 to DMEM/F12 (1:1) supplemented with 10 % of FBS, 100 U/ml of penicillin, 100 μg/ml of streptomycin.
5. PCR and DNA sequencing primers: Reconstitute the oligos to 20 pmol/μl in deionized, sterile water. Store the reconstituted oligos at −20 °C. pSM2-Xho-Sup-72 and pSM2-EHRev are the primers for amplifying shRNA-containing DNA from genomic DNA. pSM2C sequencing primer is for DNA sequencing. Oligonucleotide sequences are:

   pSM2-Xho-Sup-72: 5′-GCTCGCTTCGGCAGCACATATAC-3′.

   pSM2-EHRev: 5′-GAGACGTGCTACTTCCATTTGTC-3′.

   pSM2C sequencing primer: 5′-GAGGGCCTATTTCCCATGAT-3′.

3 Methods

3.1 Preparing an ATF5 Promoter-Driven DTR Expression Construct

1. Mix 100 ng of the plasmid MCD11C-DTR-EGFP, 20 pmol of the primers DTR-NT-RI and EGFP-N1-CT-Xba1 with overhangs containing either EcoRI or XbaI restriction enzyme sites, 10 nmol of dNTP, and 1.25 U of Taq DNA polymerase. Perform PCR using the following PCR cycle setting: 92 °C for 3 min; 92 °C for 10 s, 56 °C for 30 s, and 68 °C for 2 min (repeat 24 times).
2. Digest the PCR products and the promoter-less vector pMyc-His B with EcoRI and XbaI. Dephosphorylate the vector with CIP. Perform ligation, bacterial transformation, and clone selection using the method described in Subheading 2.1, **item 3**. Name the resulting plasmid as pDTR-EGFP (*see* **Note 9**).
3. Mix 100 ng of the plasmid pGL4.14-*Atf5p*-4 kb with 20 pmol of the primers *Atf5p*-4117 bp and *Atf5p*-Rev with overhangs containing MluI or SpeI restriction enzyme sites, 10 nmol of dNTP, and 1.25 U of Taq DNA polymerase. The PCR cycle is as follows: 93 °C for 3 min; 93 °C for 20 s, 53 °C for 30 s, and 68 °C for 4 min (repeat 29 times).
4. Clone the amplified mouse *Atf5* promoter into pDTR-EGFP using the method described in Subheading 2.1, **item 3**. Designate the resulting plasmid p*Atf5p*-DTR-EGFP (*see* **Note 10**).

3.2 Determining the DT Killing Curve

1. Seed 4×10^5 GL261 mouse malignant glioma or U251 human glioblastoma cells in a 6-well plate. Incubate at 37 °C overnight.
2. Thaw an aliquot of DT (20 μM) stock solution on ice. Prepare working solutions of DT by diluting DT (20 μM) stock solution into various concentrations using DT-dissolving buffer. Add DT working solutions (1:1,000) to GL261 culture media to make DT selection media (*see* **Note 11**).
3. Add freshly made DT selection media to the cells, and incubate at 37 °C for 7 days. Replenish cells with freshly made DT selection media every other day.
4. Remove and discard media (*see* **Note 12**). Gently add 1–2 ml of crystal violet solution (0.5 %). Incubate cells at RT for 20 min. Transfer used crystal violet solution into a separate glass bottle. Save the solution for future reuse.
5. To wash the plates fill a container with tap water (the container should be large enough to accommodate a 6-well plate). Dip the plate into the water, fill each well with the water, pour out the water, and refill it. Wash the plate several times until no crystal violet comes off.
6. Air-dry the plate by placing it face down on a paper towel. Save the results by scanning the plate using a scanner with high resolution (Figs. 1 and 2). Count the colonies with more than

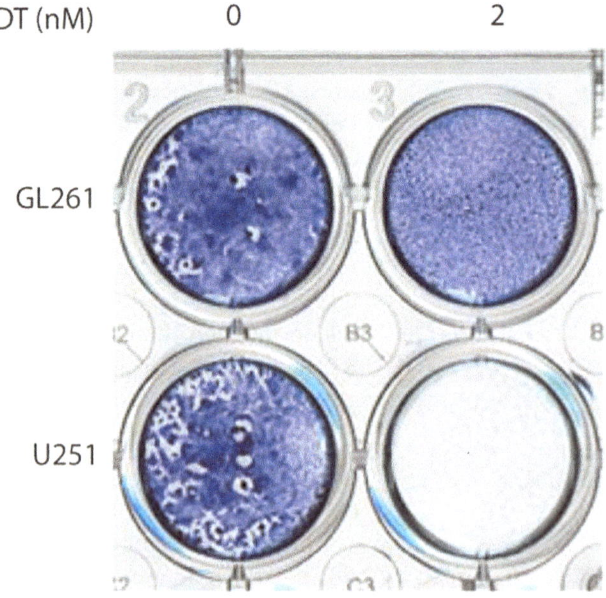

Fig. 1 DT kills human but not mouse cells. GL261 mouse malignant glioma and U251 human glioblastoma cells were plated in a 24-well plate and treated with 0 or 2 nM of DT. 7 days later, cells were stained with crystal violet

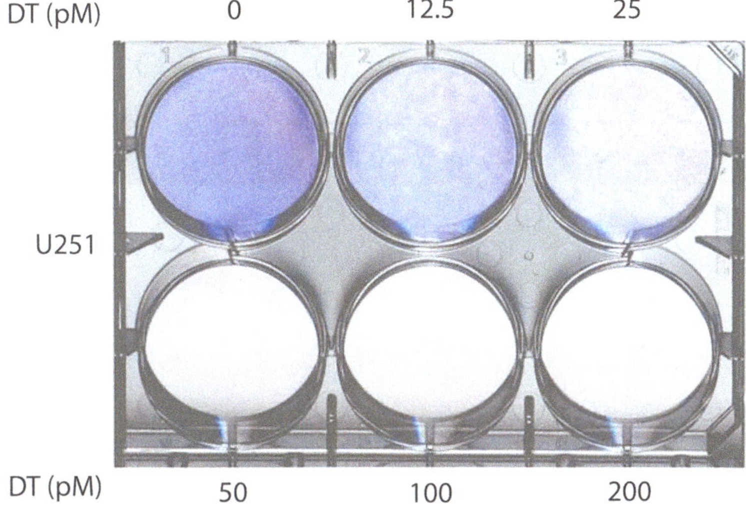

Fig. 2 Cytotoxicity of DT to U251 cells. U251 human glioblastoma cells were plated in a 6-well plate and treated with DT at concentrations ranging from 0 to 200 pM. Cells were stained with crystal violet after 7-day treatment

50 cells. No colonies should be expected at the concentrations that DT is able to kill all human cells (*see* **Note 13**). DT killing curve helps to determine the dose of DT to be used in the following experiments.

3.3 Establishing a DT-Sensitive Mouse Cell Line

1. Digest the plasmid p*Atf5p*-DTR-EGFP (Subheading 3.1) with PvuI (Promega) to linearize the plasmid (*see* **Note 14**).
2. Seed 4×10^5 GL261/pCMV-3xFLAG-ATF5 cells in G418 (0.25 mg/ml)-maintaining media in a 6-well plate. Incubate at 37 °C overnight.
3. On the next day, mix 1 μg of PvuI-digested p*Atf5*p-DTR-EGFP with 8 μl of enhancer in 100 μl of buffer EC. Vortex, and briefly spin. Incubate at RT for 5 min. Add 10 μl of Effectene (QIAGEN), and mix by pipetting. Incubate at RT for 15 min. During incubation, replenish the cells with 2 ml of fresh G418-free media. Add the transfection mix dropwise to the cells.
4. 48 h later, add 5 ml of the freshly made zeocin (50 μg/ml) selection media to the cells. Incubate at 37 °C for 15 days. Replenish the cells with freshly made zeocin selection media every 3 days.
5. Trypsinize and resuspend cells at a concentration of 1 cell per 200 μl zeocin selection media, and make enough for two 96-well plates. Add 200 μl of cell suspension per well in two 96-well plates (*see* **Note 15**). Incubate at 37 °C for 3–4 weeks. Replenish cells with freshly made zeocin selection media every week.

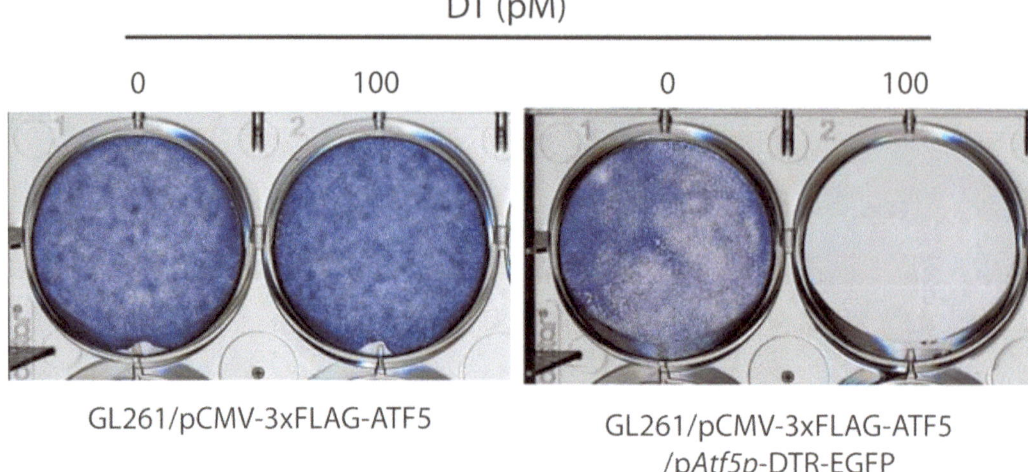

Fig. 3 Establishing a DT-sensitive mouse cell line. GL261/pCMV-3xFALG-ATF5 or GL261/pCMV-3xFLAG-ATF5/*pAtf5p*-DTR-EGFP cells were plated in 6-well plates and treated with 100 pM of DT. Cells were stained with crystal violet after 7-day treatment

6. Monitor cells under an inverted fluorescence microscope for EGFP expression. When cells become confluent in 96-well plates, transfer the EGFP+ cells to 24-well plates for expansion and maintain the cells in G418 (0.25 mg/ml)- and zeocin (20 μg/ml)-maintaining media.
7. When cells become confluent, perform DT toxicity assay to screen DT-sensitive cells. Be sure to save a part of the cells and use the rest for the following assay.
8. Inoculate 4×10^4 or 4×10^5 cells per well in G418/zeocin-maintaining media in 24-well plates or 6-well plates, respectively. Incubate at 37 °C overnight.
9. Add 5 ml of the freshly prepared DT (100 pM) selection media (*see* **Note 16**) to each well, and incubate cells at 37 °C for 7 days. Replenish cells with freshly made DT selection media every other day.
10. Remove the media, and stain the cells with crystal violet using the method described in Subheading 3.2, **steps 4–6**. Save the results (Fig. 3) by scanning the plate. The DT-sensitive cells should be killed by 100 pM of DT (*see* **Note 16**). Designate the resulting cell line GL261/pCMV-3xFLAG-ATF5/p*Atf5p*-DTR-EGFP. Maintain the cells in G418/zeocin-maintaining media.

3.4 Determining the Titer of shRNA Library Viruses

1. Seed 1×10^5 HEK293T cells in a 12-well plate (*see* **Note 17**). Incubate at 37 °C overnight.
2. Dilute the shRNA library viruses by 1:200, 1:2,000, 1:20,000, 1:200,000, 1:2,000,000, or 1:20,000,000 in 1 ml culture

media. Add 1 μl of polybrene stock solution (final concentration 10 μg/ml). Mix by vortexing. Remove the media and add the virus dilutions slowly to the cells.

3. Incubate cells at 37 °C for 24 h. Change the media to puromycin selection media (1 μg/ml). Incubate at 37 °C for 2–3 weeks. Change the media every week.
4. Remove the media, and stain the cells with crystal violet (described in Subheading 3.2, **steps 4–6**). Count colonies, and determine the virus titer using the following formula (*see* **Note 17**): Virus titer = [colony number/virus volume (μl)] × 1,000. The acceptable titer for the shRNA library viruses is at least 10^6 IU/ml.

3.5 Performing an RNAi Screen

1. For each pool of shRNAs (13 pools for the entire mouse shRNA library containing 70,000 constructs), seed 5×10^6 GL261/pCMV-3xFLAG-ATF5/pAtf5p-DTR-EGFP cells in G418/zeocin-maintaining media in a 100 mm cell culture dish. Incubate overnight at 37 °C.
2. Infect the cells at a multiplicity of infection (MOI) of 1 (*see* **Note 18**). Determine the volume (ml) of shRNA library virus using the following formula: (cell number × MOI)/virus titer. Mix 5 ml of a pool of shRNA library virus with 5 μl of polybrene stock solution (final concentration 10 μg/ml). Add the mixture to the cells carefully, and incubate cells at 37 °C for 2 h. Add another 5 ml of G418/zeocin-maintaining media, and incubate cells at 37 °C for 48 h.
3. Change the media to puromycin (1 μg/ml) selection media. Incubate cells at 37 °C for 7 days (*see* **Note 19**).
4. Change the media to freshly made DT (100 pM) selection media. Incubate cells at 37 °C for another 7 days. Replenish cells with freshly made DT selection media every other day.
5. Change the media to GL261 cell culture media without G418, zeocin, puromycin, and DT. Incubate cells at 37 °C for 2–3 weeks until colonies form (*see* **Note 20**).
6. Isolate and expand the colonies (described in Subheading 3.3, **steps 5** and **6**).
7. When the cells become 90 % confluent, trypsinize and collect cells by centrifugation at 500 × *g* for 5 min. Wash cell pellets once with ice-cold 1× PBS buffer. Isolate genomic DNA using QIAGEN DNeasy kit, and measure the DNA concentration.
8. Mix the following in a PCR reaction tube: 500 ng of genomic DNA, 10 nmol of dNTP mix, 20 pmol of pSM2-Xho-Sup-72 or pSM2-EHRev, 5 μl of 10× PCR buffer, and 1.25 U of Taq DNA polymerase. Perform the PCR in the following cycle setting: 95 °C for 5 min; 95 °C for 1 min, 60 °C for 1 min,

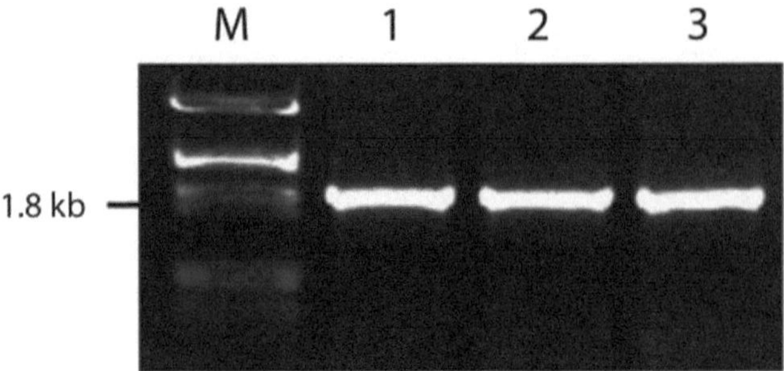

Fig. 4 Amplification of genomic DNAs containing shRNA sequences. Genomic DNAs containing shRNA sequences were amplified using primers that span the flank regions of shRNA inserts. 2-kb DNA fragments were resolved on a 1 % agarose gel. M is a DNA ladder. Samples 1–3 are from three different DT-resistant cells isolated in the screen described in Subheading 3.5

and 72 °C for 2 min (repeat 29 times); 72 °C for 30 min. Resolve the PCR products in a 1 % agarose gel (Fig. 4).

9. Mix 3 μl of the PCR product with 5 μl of 2× ligase buffer, 1 μl of pGEM-T vector, and 1 μl of T4 DNA ligase (Promega). Incubate at 4 °C overnight. Transform the ligation mixture into competent DH5α cells by heat shock. Plate cells in agar dishes with 50 μg/ml ampicillin.
10. Pick up single bacterial colonies, and prepare the plasmids using Wizard® Plus SV Minipreps DNA Purification System (Promega). Send the purified plasmids for regulator DNA sequencing using pSM2C sequencing primer.
11. Identify shRNA sequences using XhoI (CTCGAG) and EcoRI (GAATTC), two restriction enzymes that span the flanking region of 97 bp shRNA insert. Copy and paste the shRNA sequence into the online database RNAi codex, and search for the identity of each individual shRNA. Validate the shRNA identity using a different online database: NCBI blast.
12. Repeat **steps 10** and **11** until many shRNA candidates are repeatedly sequenced (*see* **Note 21**).

4 Notes

1. This promoter-less vector is designed for expressing reporter genes driven by the promoter of interest. Inserting a DNA sequence containing multi-cloning sites facilitates future cloning.
2. ATF5 is essential for the survival of GL261 mouse malignant glioma cells [10]. Thus, knocking down ATF5 upstream transcriptional activators inhibits cell survival. Expression of

an exogenous ATF5 compensates for the loss of the endogenous one. This strategy is useful for studying genes essential for cell survival.

3. Reconstituted DT is stable if stored at −80 °C. However, it is labile during freeze/thaw cycle. Thus, to avoid significant loss of activity, it is better to make a DT stock solution at high concentrations and make aliquots in small amount for one-time use.
4. Loss of expression of exogenously inserted genes is a problem that occurs frequently when making stable cell lines. We find that maintaining the established cell lines in media containing a lower dose of antibiotics helps stabilize the expression of inserted genes.
5. HEK293T cells easily detach when they become confluent. Although this may be prevented by seeding fewer cells, the virus titer will be lower. We find that HEK293T grown in poly-D-lysine-coated cell culture plates do not detach when confluent. To coat plates with poly-D-lysine, add poly-D-lysine (50 μg/ml) to each well to cover the surface. Incubate at 37 °C for 30 min. Aspirate the excess poly-D-lysine, and leave plates uncovered in a biosafety cabinet until completely dry. The plates may be used immediately or stored at 4 °C.
6. The shRNA library from Thermo Scientific/Open Biosystems includes about 70,000 constructs divided into 13 pools. The constructs and viruses are prepared in the RNAi core facility at the University of Massachusetts Medical School. The plasmid mix and the viruses of shRNA libraries are also commercially available.
7. The ratio of the plasmids for producing viruses is critical. We find that transfecting HEK293T cells with shRNA vector, psPAX2, and pMD2.g at a ratio of 1:1:0.5 produces the viruses with a titer at least 10^6 IU/ml.
8. Freshly made viruses can be used directly to infect cells or stored at 4 °C without significantly losing transduction efficiency. However, we find that it is better to aliquot the viruses and keep them at −80 °C for long-term storage. Avoid frequent freeze/thaw as it lowers transduction efficiency. Filtering the viral supernatant is necessary to keep it sterile and free of HEK293T cells.
9. Preparing a construct encoding a protein fused with GFP facilitates the evaluation of the fusion protein when stably expressed in cells and also the activity of the promoter that drives its expression. Our study, together with others, demonstrated that DTR fusion protein retains its biological activity in responding to DT [9, 10]. However, for the purpose of RNAi screen, it is not necessary to make a DTR fusion protein.

10. pDTR-EGFP is a promoter-less vector, which can be used to study any promoter of interest. It is a useful tool to study trans-factors that control the activity of DNA cis-elements. The length of DNA upstream of the transcription start site is optional depending on the RNAi screen settings. We find that it is better to use a longer promoter as many trans-factors bind to the regions far away from transcription start site.
11. When studying the DT killing at a series of concentrations, we find that it is convenient to make DT working solutions at high concentrations and then dilute 1:1,000 to obtain the final concentrations. This way, an equal amount of DT-dissolving buffer is added to each sample increasing the accuracy of DT treatment.
12. GL261 cells grow normally as adherent cells, and they are dying when they detach. Therefore it is critical to remove the media and floating dead cells as much as possible.
13. DT killing curve is critical for determining the concentration of DT to be used in RNAi screening. Results in Fig. 2 show that as low as 50 pM DT kills all human U251 cells.
14. We find that linearizing the plasmid increases the efficiency of making a stable cell line.
15. We find that diluting cells and plating them in 96-well plates facilitate the drug selection in making stable cell lines. However, it can also be done by plating diluted cells (usually 100–1,000 cells per 10 ml of media) in 100 mm dishes and waiting for colonies to form.
16. Based on the results in Fig. 2, we used 100 pM DT to select GL261 cells that became sensitive to DT upon expression of human DTR (Subheading 3.3) and to select DT-resistant cells in an RNAi screen (Subheading 3.5). The purpose of using a higher concentration of DT (*see* **Note 13**) is to decrease the false-positive candidates when making stable cell lines or performing an RNAi screen (*see* **Note 21**).
17. We find that a virus dilution that gives rise to 100–500 colonies yields accurate virus titer. Balancing the cell number plated initially and the virus dilutions helps to determine the virus titers.
18. When transducing cells with viruses containing shRNAs, it is possible that one cell may take up two or more shRNAs, which compromises RNAi screening. Previous results showed that transducing cells at an MOI of 0.2–2 limits such possibility [3]. We empirically determined the MOI used to infect GL261 cells in this study.
19. Puromycin selection is to ensure the expression of shRNA in infected cells. We find that it is unnecessary to treat cells with puromycin for longer than 7 days in our RNAi screen because

puromycin rapidly kills cells with no expression of puromycin-resistant gene Pac (puromycin *N*-acetyl-transferase). Plus, it is unnecessary to establish cell lines stably expressing a library of shRNAs when performing an RNAi screen. In order not to increase the unnecessary selection pressure, we intentionally remove G418 and zeocin from puromycin selection media (Subheading 2.4).

20. This step is essential for the recovery of surviving cells after the selection of puromycin and DT. However, it also increases false-positive candidates in the screen. We find that it is necessary to perform a secondary screen to validate the candidate shRNAs, which is not described in this chapter.

21. We find that it is necessary to exhaust DNA sequencing. In this screen, we have submitted 300 samples for DNA sequencing, in which 61 unique shRNAs were identified. 22 shRNAs were found to have at least 2 repeats in DNA sequencing. Thus, it is important to sequence at least 300–500 samples in order to acquire a full list of candidate shRNAs. The number of samples to be sequenced largely depends on the screening settings. Increasing the stringency in DT selection will significantly eliminate the false-positive candidates and diminishes the demand of sequencing large number of samples.

Acknowledgement

We thank Claude Gazin and Amy Virbasius in assisting with experiments. This work was supported by the start-up funds from Virginia Tech Carilion Research Institute to Z.S.

References

1. Paddison PJ, Silva JM, Conklin DS, Schlabach M, Li M, Aruleba S, Balija V, O'Shaughnessy A, Gnoj L, Scobie K, Chang K, Westbrook T, Cleary M, Sachidanandam R, McCombie WR, Elledge SJ, Hannon GJ (2004) A resource for large-scale RNA-interference-based screens in mammals. Nature 428:427–431
2. Chang K, Elledge SJ, Hannon GJ (2006) Lessons from Nature: microRNA-based shRNA libraries. Nat Methods 3:707–714
3. Gazin C, Wajapeyee N, Gobeil S, Virbasius CM, Green MR (2007) An elaborate pathway required for Ras-mediated epigenetic silencing. Nature 449:1073–1077
4. Campisi J (2005) Suppressing cancer: the importance of being senescent. Science 309:886–887
5. Wajapeyee N, Serra RW, Zhu X, Mahalingam M, Green MR (2008) Oncogenic BRAF induces senescence and apoptosis through pathways mediated by the secreted protein IGFBP7. Cell 132:363–374
6. Greene LA, Lee HY, Angelastro JM (2009) The transcription factor ATF5: role in neurodevelopment and neural tumors. J Neurochem 108:11–22
7. Monaco SE, Angelastro JM, Szabolcs M, Greene LA (2007) The transcription factor ATF5 is widely expressed in carcinomas, and interference with its function selectively kills neoplastic, but not nontransformed, breast cell lines. Int J Cancer 120:1883–1890
8. Collier RJ (1975) Diphtheria toxin: mode of action and structure. Bacteriol Rev 39:54–85

9. Jung S, Unutmaz D, Wong P, Sano G, De los Santos K, Sparwasser T, Wu S, Vuthoori S, Ko K, Zavala F, Pamer EG, Littman DR, Lang RA (2002) In vivo depletion of CD11c+ dendritic cells abrogates priming of CD8+ T cells by exogenous cell-associated antigens. Immunity 17:211–220

10. Sheng Z, Li L, Zhu LJ, Smith TW, Demers A, Ross AH, Moser RP, Green MR (2010) A genome-wide RNA interference screen reveals an essential CREB3L2-ATF5-MCL1 survival pathway in malignant glioma with therapeutic implications. Nat Med 16: 671–677

Chapter 7

Cancer Metabolism: Cross Talk Between Signaling and O-GlcNAcylation

Christina M. Ferrer and Mauricio J. Reginato

Abstract

Cancer cells exhibit a unique metabolic shift to aerobic glycolysis that has been exploited diagnostically and therapeutically in the clinic. Oncogenes and tumor suppressors alter signaling pathways that lead to alterations of glycolytic flux. Stemming from glycolysis, the hexosamine biosynthetic pathway leads to elevated posttranslational addition of O-linked-β-*N*-acetylglucosamine (O-GlcNAc) on a diverse population of nuclear and cytosolic proteins, many of which regulate signaling pathways. This unit outlines techniques used to detect metabolic alterations in cancer cells, regulation by signaling pathways, and cellular O-GlcNAcylation.

Key words Cancer, Metabolism, Signaling, O-GlcNAcylation, O-GlcNAc transferase, Glucose uptake, Lactate, ATP, Hexosamine biosynthetic pathway, mTOR, LKB1, AMPK, Hypoxia, HIF-1α

1 Introduction

To support rapid growth cancer cells display altered metabolic needs that result in a production of ATP through aerobic glycolysis even in the presence of normal oxygen levels. The resulting metabolic shift to increase biomass through increased glycolytic flux is termed the "Warburg effect" and is essential for supporting oncogenic phenotypes by regulating cancer cell growth and survival [1, 2].

This enhanced glucose uptake provides additional glycolytic intermediates which are available to enter nutrient signaling pathways such as the hexosamine biosynthetic pathway (HBP) [3, 4]. While the majority of glucose that enters the cell is used in the glycolytic pathway, 2–5 % of this glucose is diverted to the HBP. The terminal product of the HBP is the activated intermediate UDP-GlcNAc, which can be conjugated to serine and threonine residues of nuclear and cytosolic proteins through the action of a single enzyme O-linked-β-*N*-acetylglucosamine (O-GlcNAc) transferase (OGT). This enzyme is the sole known glycosyltransferase

Narendra Wajapeyee (ed.), *Cancer Genomics and Proteomics: Methods and Protocols*, Methods in Molecular Biology, vol. 1176, DOI 10.1007/978-1-4939-0992-6_7, © Springer Science+Business Media New York 2014

responsible for this posttranslational modification, and this modification can be removed by the glycoside hydrolase O-GlcNAcase (OGA) that catalyzes cleavage of O-GlcNAc from proteins. This nutrient-sensitive O-GlcNAc modification modulates protein dynamics as well as widespread cellular signaling pathways in mammalian cells. Modulation of O-GlcNAc levels has been linked to altered cellular development, mitotic progression, and growth and survival patterns [5]. Most importantly, we and other groups have shown that O-GlcNAcylation can regulate cancer phenotypes both in vitro and in vivo in part through regulation of key oncogenic factors [6, 7] and proteins central to the Warburg effect [8, 9].

Oncogene and tumor-suppressor genes have been shown to directly regulate enzymes critical for glycolysis in cancer cells. A major oncogenic pathway that is altered in 80 % of all cancers is the receptor tyrosine kinase (RTK)/PI3 kinase/mammalian target of rapamycin (mTOR) pathway [10]. This pathway is central to shift in metabolism as it can regulate translation of key transcription factors including HIF-1α and c-Myc that directly regulate expression of many, if not all, glycolytic enzymes [10]. Multiple environmental cues, including nutrients, can regulate mTOR signaling including the tumor suppressor LKB1, which activates AMPK. This activation of AMPK leads to inhibition of mTOR activity and loss of mTOR signaling, in turn, resulting in a decrease in the translation of critical cell growth and metabolic regulators [11]. Cancer cells often outgrow their oxygen supply and become hypoxic which leads to further stabilization of HIF-1α levels and further drives cancer cells into glycolysis [12]. Emerging data from our lab and others suggests that the nutrient sensor O-GlcNAcylation can link signaling pathways to metabolic regulation in cancer cells. Here, we describe methods to examine glycolytic flux, metabolic associated signaling under normoxic and hypoxic conditions, and O-GlcNAcylation in cancer cells. We combine basic techniques to analyze glycolytic metabolites with detection of metabolic signaling pathways and their interplay with O-GlcNAcylation to understand cross talk between these pathways in regulating transformation and cancer progression. Cross talk between oncogenes and metabolic pathways is critical for understanding cancer cell transformation and malignant phenotypes that may allow us to specifically target these pathways therapeutically.

2 Materials

Prepare and store all reagents as per the manufacturer's instructions.

2.1 Fluorescently Labeled Glucose Uptake Assay

1. 2-NBD glucose.
2. Trypsin/EDTA 1×: 0.25 % Trypsin/2.21 mM EDTA in HBSS without sodium bicarbonate, calcium, and magnesium.

3. Phosphate-buffered saline (PBS) 1×.
4. Flow cytometer with the ability to detect fluorescently labeled cells.

2.2 Lactate Production Assay

1. Lactate Colorimetric Assay Kit II (BioVision K627-100).
2. Trypsin/EDTA 1×: 0.25 % Trypsin/2.21 mM EDTA in HBSS without sodium bicarbonate, calcium, and magnesium.
3. PBS 1×.
4. Polystyrene flat-bottom 96-well microplate without lid.
5. Spectrophotometer or microplate reader able to detect absorbance at 450 nm.

2.3 ATP Quantitation Assay

1. Adenosine 5′-triphosphate (ATP) Bioluminescent Assay Kit (Sigma-Aldrich FLAA-1KT).
2. Trypsin/EDTA 1×: 0.25 % Trypsin/2.21 mM EDTA in HBSS without sodium bicarbonate, calcium, and magnesium.
3. PBS 1×.
4. Black polystyrene flat-bottom 96-well microplate without lid.
5. Luminometer.

2.4 Antibodies

1. Sensing Cellular Energy Status
   (a) Anti-LKB1 (Cell Signaling (CS), 27D10), anti-phosphoserine 428 LKB1 (CS, 3482), anti-AMP kinase (CS, 5831), anti-phosphothreonine 172 AMP kinase (CS, 5256), anti-phosphoserine 79-acetyl CoA carboxylase 1 (ACC) (CS, 3661), anti-ACC (CS, 3676).
2. Growth Factor and Metabolic Signaling Pathways
   (a) Anti-phospho-HER2/ErbB2 (CS, 6942), anti-HER2/ErbB2 (CS, 2164), anti-phospho-EGFR (CS, 2235), anti-EGFR (CS, 4267).
   (b) Anti-phospho-AKT (Ser473) (CS, 4070), anti-phospho-AKT (Thr308) (CS, 4056), anti-AKT (CS, 4691), anti-phospho-p44/42 MAPK(Erk1/2) (Thr202/Tyr204) (Santa Cruz, SC-16982), total p44/42 MAPK (Santa Cruz, SC-1647).
   (c) Anti-Tuberin/TSC2 (CS, 3612), anti-phospho-Raptor (Ser792) (CS, 2083), anti-Raptor (CS, 2280), anti-p70 S6 kinase (CS, 2708), anti-phospho-S6 kinase (Thr389) (CS, 9202), anti-phospho-4EBP1 (Thr37/46) (CS, 3929), anti-4EBP1 (CS, 9466).
3. Transcription Factors That Promote the Warburg Effect: Hypoxia-Inducible Factors
   (a) Anti-HIF-1α (Novus Biologicals), anti-HIF-1α (immunoprecipitations) (Abcam H1alpha67), anti-HIF-1α (hydroxy

P564) (Abcam), anti-K48 linkage-specific polyubiquitin antibody (CS, 12805), anti-pVHL (Novus Biologicals).

4. HBP and O-GlcNAcylation

   Detecting Pathway Enzyme Expression

   (a) Anti-GFAT (CS, 5322), anti-OGT (Sigma), anti-OGA (MGEA5) (Abcam).

   Detecting Global O-GlcNAcylation

   (b) Anti-O-GlcNAc (RL2) (Santa Cruz), anti-O-GlcNAc (CTD110.6) (Sigma).

2.5 Pharmacological Inhibitors and Activators

Note: Dissolve all chemicals as per the manufacturer's instructions.

1. Phenformin HCl.
2. 5-Aminoimidazole-4-carboxamide-ribonucleoside—AICAR.
3. Compound C—6-[4-(2-Piperidin-1-ylethoxy)phenyl]-3-pyridin-4-ylpyrazolo[1,5-a]pyrimidine.
4. Lapatinib Ditosylate—Tykerb.
5. AG1478—Tyrophostin.
6. LY-294002 hydrochloride.
7. BEZ-235.
8. Rapamycin.
9. Dimethyloxaloylglycine—DMOG.
10. $CoCl_2$—Cobalt (II) chloride hexahydrate.
11. Lactacystin.
12. OGA inhibitor: 6-Acetamido-6-deoxy-castanospermine (6-Ac-Cas)—(Dr. David J. Vocadlo, Simon Fraser University).
13. OGT inhibitor: Ac-5SGlcNAc—(Dr. David J. Vocadlo, Simon Fraser University).

2.6 Overexpression Plasmids

1. pBabe-puromycin-ErbB2 (NeuT)—Plasmid overexpressing constitutively active form of ErbB2 [13].
2. pBabe-puromycin-Myr-AKT1—Plasmid overexpressing myristoylated form of AKT1 (active AKT1) [14].
3. pMIT-HIF-1α-P402/564A mutant [15].
4. pBabe-puro-VHL [16].

2.7 Immortalized and Cancer Cell Lines

1. MCF-10A cells (ATCC, CAT. CRL-10317).
2. MDA-MB-231 cells (ATCC, Cat. HTB-26).
3. PC3 cells (ATCC, Cat. CRL-1435).

2.8 Hypoxia

Ruskinn $Invivo_2$ 400 Low Oxygen Workstation.

2.9 QRT-PCR

1. RNeasy RNA Extraction Kit (Qiagen, Cat. 74104).
2. Brilliant II QRT-PCR Master Mix Kit (Stratagene, Cat. 600809).
3. qPCR tubes (VWR, Cat. 99900-144).
4. Primer/probe sets—Applied Biosystems (TaqMan).

   OGT—Cat. HS00914634_gl.

   HIF-1α—HS00936371_ml.

   HIF-1α transcription targets:

   ADM—HS00181605_ml.

   GLUT1—HS00892681_ml.

   LDHA—HS00855332_gl.

   BNIP3L—HS00108949_ml.

3 Methods

Assays described below have been used in cancer cell lines (such as breast cancer cells MDA-MB-231 or prostate cancer cells PC3) to determine changes in signaling and/or metabolism under different conditions. In addition, we have also used immortalized human mammary epithelial cells (such as MCF-10A) overexpressing a specific oncogene (such as ErbB2 or Akt). This is an excellent cell system to understand direct regulation of metabolic outputs including O-GlcNAcylation by specific oncogenes. Methods for culturing MDA-MB-231 [6], PC3 [7], and MCF-10A overexpressing oncogenes [14, 17] have been previously described.

3.1 Measuring Glycolytic Flux

This assay detects the uptake of a fluorescent D-glucose analog via a flow cytometric method in living cells.

3.1.1 Fluorescently Labeled Glucose Uptake Assay

1. Plate cells (*see* **Note 1**), and following cell treatment (with RNAi or inhibitor) change cell media and add 2-NBD glucose at a final dilution of 1:1,000 (*see* **Note 2**).
2. Incubate at 37 °C for 2 h.
3. Aspirate media incubated with 2-NBD glucose, and add trypsin.
4. Collect cell pellets and wash twice with 1× PBS.
5. Resuspend cell pellets in the appropriate volume of 1× PBS for flow cytometry evaluation of fluorescently labeled cells (*see* **Note 3**).

3.1.2 Lactate Production Assay

In this assay, lactate is oxidized by lactate dehydrogenase to generate a product which interacts with a probe to produce a color (λ_{max} = 450 nm). Assay detects 0.02–10 mM lactate levels.

1. After treatment, collect 200 μl of *culture media* (*see* **Note 4**).
2. Trypsinize cells, collect cell pellets, and wash twice with 1× PBS.
3. Add 1–50 μl of *cell culture media* from **step 1** to 96-well plate, and adjust to a final volume of 50 μl with assay buffer.
4. Prepare D-Lactate Standard curve dilutions. Dilute the 100 mM D-Lactate Standard to 1 mM by adding 10 μl of the standard to 990 μl of assay buffer, and mix well. Add 0, 2, 4, 6, 8, and 10 μl into a series of wells. Adjust volume to 50 μl/well with assay buffer to generate 0, 2, 4, 6, 8, and 10 nmol/well of the D-Lactate Standard.
5. For each well, prepare a total 50 μl reaction mix containing the following components:

   46 μl of D-Lactate assay buffer

   2 μl of D-Lactate substrate mix

   2 μl of D-Lactate enzyme mix
6. Add 50 μl of the reaction mix to each well containing the D-Lactate Standard or test samples, and mix well.
7. Incubate for 30 min at room temperature.
8. Measure OD 450 nm in a microplate reader.
9. Correct background by subtracting the value derived from the 0 D-lactate control from all standard and sample readings (*see* **Note 5**).
10. Collect cell lysates from **step 3**, and normalize values obtained in **step 9** to total the amount of protein in each sample.

3.2 Sensing Cellular Energy Status

3.2.1 ATP Production Assay

1. Prepare reagents:
   (a) Reconstitute ATP monitoring enzyme (*see* **Note 6**) with 220 μl of the enzyme reconstitution buffer. Mix gently by inversion (do not vortex).
   (b) Aliquot enough enzyme (1 μl per assay) for the number of assays to be performed in each experiment and freeze at −80 °C for future use.
   (c) Prepare ATP standard solution by dissolving the 1 mg ATP into 1 ml dH_2O.
2. Add 100 μl of nucleotide-releasing buffer to the wells that will contain samples.
3. Plate cells (*see* **Note 1**), following treatment, trypsinize cells, resuspend in PBS, and transfer 10 μl of cultured cells (containing 10^3–10^4 cells) onto the wells of a flat-bottom black 96-well plate.
4. Cover, and incubate at room temperature for 5 min on a gentle rocker.

5. Prepare a standard curve with ATP provided in the kit:

ng ATP	µl ATP Solution	µl H_2O
100	13 µl Stock	117
10	100 µl of 100 ng	1,000
1	10 µl of 10 ng	100
0.1	100 µl of 1 ng	1,000
0.01	10 µl of 0.1 ng	100
0	0	0

6. Add 100 µl nucleotide-releasing buffer in the 6 wells containing the standards.
7. Add 10 µl of each standard to the wells in **step 6** and 10 µl nucleotide-releasing buffer to the well containing 0 ng ATP.
8. Add 10 µl of ATP monitoring enzyme dilution to each well containing samples and standards (*see* **Note 7**).
9. Read plate on a luminometer expressing ATP levels relative to control values for relative ATP levels. To calculate unknown absolute ATP levels, graph and fit the logarithmic trend line.

3.2.2 Using Antibodies to Detect Cellular Energy Status

1. Prepare cell lysates from 1 to 5×10^6 cells in radioimmune precipitation assay buffer (150 mM NaCl, 1 % NP40, 0.5 % DOC, 50 mM Tris–HCl at pH 8, 0.1 % SDS, 10 % glycerol, 5 mM EDTA, 20 mM NaF, and 1 mM Na_3VO_4) supplemented with 1 µg/ml each of pepstatin, leupeptin, aprotinin, and 200 µg/ml PMSF.
2. Clear lysates by centrifugation at $16,000 \times g$ for 15 min at 4 °C and analyze by SDS-PAGE and autoradiography.
3. Visualize proteins by immunoblotting using primary antibodies indicated in Subheading 2.4.

3.2.3 Inhibitors and Activators of AMPK

A. Activation of AMPK

Phenformin

1. Seed 2.5×10^5 cells 24 h prior to treatment with phenformin.
2. Treat cells with 50–5,000 µM final concentration of phenformin in cell culture media for 24 h.
3. Collect lysates using appropriate methods, and perform Western blot for phosphorylation of AMPK at Thr172, total AMPK levels, and phosphorylation of ACC at Ser79 and total levels of ACC.

AICAR

1. Seed 2.5×10^5 cells 24 h prior to treatment with AICAR.
2. Treat cells with 10–300 μM final concentration of AICAR in cell culture media for 24 h.
3. Collect lysates using appropriate methods, and perform Western blot for phosphorylation of AMPK at Thr172 and total AMPK levels and phosphorylation of ACC at ser79 and total levels of ACC.

B. Inhibition of AMPK Activity

Compound C—Commonly used for studying AMPK-dependent cellular events in vitro.

1. Seed 2.5×10^5 cells 24 h prior to treatment with compound C.
2. Treat cells with final concentration of 10 μM compound C in cell culture media for 24 h.
3. Collect lysates using appropriate methods, and perform Western blot for inhibition of phosphorylation of AMPK target, acetyl CoA carboxylase (p-ACC).

3.3 Oncogenes and Tumor Suppressors: Regulators of Cancer Metabolism

3.3.1 Activation and Detection of Growth Factor Signaling

A. Using Antibodies to Detect Changes in Cellular Signaling

1. Cell lysates from 1 to 5×10^6 cells were prepared in radio-immune precipitation assay buffer (150 mM NaCl, 1 % NP40, 0.5 % DOC, 50 mM Tris–HCl at pH 8, 0.1 % SDS, 10 % glycerol, 5 mM EDTA, 20 mM NaF, and 1 mM Na_3VO_4) supplemented with 1 μg/ml each of pepstatin, leupeptin, aprotinin, and 200 μg/ml PMSF.
2. Lysates were cleared by centrifugation at $16,000 \times g$ for 15 min at 4 °C and analyzed by SDS-PAGE and autoradiography.
3. Visualize proteins by immunoblotting using primary antibodies indicated in Subheading 2.4 (*see* **Note 8**).

B. Inhibitors and Activators of Oncogenic Receptor Tyrosine Kinases

Inhibitors of Oncogenic Receptor Tyrosine Kinases

Lapatinib ditosylate—Dual ErbB2/EGFR inhibitor

1. Seed 2.5×10^5 cells 24 h prior to treatment with lapatinib.
2. Treat cells with 10–5,000 nM final concentration of lapatinib in cell culture media for 24 h.
3. Collect lysates using appropriate methods, and perform Western blot for autophosphorylation sites of EGFR and/or ErbB2 depending on the cell line receptor expression.

AG1478—Epidermal growth factor receptor (EGFR) inhibitor

1. Seed 2.5×10^5 cells 24 h prior to treatment with AG1478.
2. Treat cells with 10–5,000 nM final concentration of AG1478 in cell culture media for 24 h.
3. Collect lysates using appropriate methods, and perform Western blot for autophosphorylation sites on EGFR.

C. Inhibitors of the PI3K/AKT Signaling Pathway

LY-294002—PI3K/AKT inhibitor

1. Seed 2.5×10^5 cells 24 h prior to treatment with LY-294002.
2. Treat cells with 5–50 μM final concentration of LY-294002 in cell culture media for 24 h.
3. Collect lysates using appropriate methods, and perform Western blot for phosphorylation of AKT.

BEZ-235—Dual ATP-competitive PI3K and mTOR inhibitor

1. Seed 2.5×10^5 cells 24 h prior to treatment with BEZ-235.
2. Treat cells with 10–5,000 nM final concentration of BEZ-235 in cell culture media for 24 h.
3. Collect lysates using appropriate methods, and perform Western blot for phosphorylation of AKT, 4EBP, or p70 S6 kinase.

3.3.2 Master Regulator of Cancer Cell Metabolism: The mTOR Pathway

A. Using Antibodies to Detect mTOR Activation

1. Prepare cell lysates from 1 to 5×10^6 cells in radioimmune precipitation assay buffer (150 mM NaCl, 1 % NP40, 0.5 % DOC, 50 mM Tris–HCl at pH 8, 0.1 % SDS, 10 % glycerol, 5 mM EDTA, 20 mM NaF, and 1 mM Na_3VO_4) supplemented with 1 μg/ml each of pepstatin, leupeptin, aprotinin, and 200 μg/ml PMSF.
2. Clear lysates by centrifugation at $16{,}000 \times g$ for 15 min at 4 °C and analyze by SDS-PAGE and autoradiography.
3. Visualize proteins by immunoblotting using primary antibodies indicated above.

B. Inhibitors of the mTOR Pathway

Rapamycin—A macrocyclic triene antibiotic that binds to and inhibits the mTOR.

1. Seed 2.5×10^5 cells 24 h prior to treatment with rapamycin.
2. Treat cells with 1–100 nM final concentration of rapamycin in cell culture media for 24 h.

3. Collect lysates using appropriate methods, and perform Western blot for phosphorylation of p70 S6 kinase and 4EBP1 using indicated antibodies.

3.4 Hypoxia and Metabolic Transcription Factors Promote the Warburg Effect

The reduction in oxygen levels in organs, tissues, or cells below 5 % is termed hypoxia. This physiologic or pathologic condition results in the stabilization of hypoxia-inducible factor (HIF) transcription factors that allow cells to adapt to reduced oxygen levels by directly regulating genes involved in angiogenesis, survival, and metabolic processes [18]. HIF is a major regulator of glycolysis in cancer cells [19]; therefore, the detection of HIF, primarily HIF-1α, is routinely used as a marker for hypoxia and/or metabolic reprogramming in cancer cells. HIF-1α protein levels are dynamically regulated by prolyl hydroxylases that regulate HIF-1α proteasomal degradation [20]. Some cancer cells contain elevated HIF-1α protein levels due to high mTOR activity or deletion of tumor suppressor VHL [10].

A. Use of Hypoxia Chamber or Hypoxic Mimetics

Hypoxia Incubator Chamber—Using the Ruskinn Invivo$_2$ 400 Low Oxygen Workstation

1. To create hypoxic conditions, adjust hypoxia chamber setting to 1 % oxygen, 5 % CO_2, and the remainder atmospheric N_2.
2. Remove most if not all oxygen present in the chamber and in your media by purging the hypoxia chamber.
3. Place the cell culture plate in the hypoxic chamber and the identical "control" plate in a normoxic incubator.
4. Typically most cells will stabilize HIF-1α protein levels in 6 h following hypoxic treatment [21]. A time course is recommended (0, 6, 12, 24 h) for individual cell line to define initial HIF-1α induction and stabilization.

DMOG—A competitive inhibitor of prolyl hydroxylase domain-containing proteins

1. Seed 2.5×10^5 cells 24 h prior to treatment with DMOG
2. Treat cells with 1 mM final concentration of DMOG in cell culture media for 24 h.
3. Collect lysates using appropriate methods, and perform Western blot analysis for HIF-1α using indicated antibodies.

$CoCl_2$—A chemical inducer of hypoxia

1. Seed 2.5×10^5 cells 24 h prior to treatment with CoCl2.
2. Treat cells with 100 μM final concentration of CoCl2 in cell culture media for 24 h.
3. Collect lysates using appropriate methods, and perform Western blot analysis for HIF-1α using indicated antibodies.

Lactacystin—Irreversible proteasome inhibitor, stabilizing HIF-1α levels

1. Seed 2.5×10^5 cells 24 h prior to treatment with lactacystin.
2. Treat cells with 5 μM final concentration of lactacystin in cell culture media for 24 h.
3. Collect lysates using appropriate methods, and perform Western blot analysis for HIF-1α using indicated antibodies.

B. Using Antibodies to Detect HIF-1α

1. Prepare cell lysates from 1 to 5×10^6 cells in radioimmune precipitation assay buffer (150 mM NaCl, 1 % NP40, 0.5 % DOC, 50 mM Tris–HCl at pH 8, 0.1 % SDS, 10 % glycerol, 5 mM EDTA, 20 mM NaF, and 1 mM Na_3VO_4) supplemented with 1 μg/ml each of pepstatin, leupeptin, aprotinin, and 200 μg/ml PMSF *in hypoxia chamber or within 10 min of removal from hypoxia chamber*.
2. Clear lysates by centrifugation at 16,000 × g for 15 min at 4 °C and analyze by SDS-PAGE and autoradiography.
3. Visualize proteins by immunoblotting using primary antibodies indicated above (*see* **Note 9**).

C. Determining HIF-1α Regulation: Posttranslational Modifications and Interaction with VHL

1. Lyse cells subjected to normoxic (21 % O_2) or hypoxic (1 % O_2) conditions (where indicated) with radioimmune precipitation assay buffer (150 mM NaCl, 1 % NP40, 0.5 % DOC, 50 mM Tris–HCl at pH 8, 0.1 % SDS, 10 % glycerol, 5 mM EDTA, 20 mM NaF, and 1 mM Na_3VO_4) supplemented with 1 μg/ml each of pepstatin, leupeptin, aprotinin, and 200 μg/ml PMSF.
2. Incubate lysates obtained in **step C1** with HIF-1α antibody (or anti-VHL antibody to detect HIF-1α/VHL interaction or HIF-OH antibody to determine the level of HIF hydroxylation) overnight at 4 °C.
3. Pre-clear Protein G Sepharose beads in 1 % bovine serum albumin (BSA) for 2 h at 4 °C on end-over-end rotator.
4. The next day, subject samples to immunoprecipitation using previously cleared Protein G Sepharose beads followed by 3× washes in 1 % PBS + Tween 20.
5. Resolve immunoprecipitated proteins by SDS-PAGE and transfer to a PVDF membrane.
6. Follow immunoblotting protocol as per the instructions above.

D. Detecting Expression of HIF-1α Targets Using Real-Time PCR

1. Isolate RNA according to RNeasy kit (Qiagen) instructions (*see* **Note 10**).

2. Determine RNA concentration.
3. Make 125 ng/μl dilution of each RNA sample in DEPC-treated H_2O.
4. Make dilution of reference dye: 0.5 μl Reference dye stock in 250 μl DEPC-treated H_2O.
5. Make a reaction mixture (components in the Stratagene kit) for each primer/probe set:

1× Reaction:	
8.75 μl	DEPC-treated H_2O
12.5 μl	Master mix
1.25 μl	Primer/probe
0.38 μl	Ref dye dilution
0.1 μl	Reverse transcriptase

6. For duplicates: Add 46 μl of the reaction mixture to a fresh microtube, and add 4 μl of each RNA dilution.
7. Flick the tube, and then spin down.
8. Pipette 25 μl of the reaction mixture with RNA into PCR tube strips, and spin down.
9. Set PCR program as follows:

   50 °C 30 min, 1 cycle

   95 °C 10 min, 1 cycle

   95 °C 15 s → 60 °C 30 s → 72 °C 30 s, 40 cycles

   72 °C 10 min, 1 cycle

3.5 Hexosamine Biosynthetic Pathway and O-GlcNAcylation

3.5.1 Detecting Pathway Enzyme Expression and Global O-GlcNAcylation

A. Using Antibodies to Detect HBP Enzyme Levels and O-GlcNAcylation

1. Prepare cell lysates from 1 to 5×10^6 cells in radioimmune precipitation assay buffer (150 mM NaCl, 1 % NP40, 0.5 % DOC, 50 mM Tris–HCl at pH 8, 0.1 % SDS, 10 % glycerol, 5 mM EDTA, 20 mM NaF, and 1 mM Na_3VO_4) supplemented with 1 μg/ml each of pepstatin, leupeptin, aprotinin, and 200 μg/ml PMSF.
2. Clear lysates by centrifugation at 16,000 × g for 15 min at 4 °C and analyze by SDS-PAGE and autoradiography.
3. Visualize proteins by immunoblotting using primary antibodies including GFAT, OGT, OGA, and O-GlcNAc.

B. Detecting Expression of O-GlcNAc Transferase Using Real-Time PCR

1. Isolate RNA according to RNeasy kit (Qiagen) instructions.
2. Determine RNA concentration.

3. Make 125 ng/μl dilution of each RNA sample in DEPC-treated H_2O.
4. Make dilution of reference dye: 0.5 μl Reference dye stock in 250 μl DEPC-treated H_2O.
5. Make a reaction mixture (components in the Stratagene kit) for OGT primer/probe set as well as control primer/probe set:

1× Reaction:	
8.75 μ	DEPC-treated H_2O
12.5 μl	Master mix
1.25 μl	Primer/probe
0.38 μl	Ref dye dilution
0.1 μl	Reverse transcriptase

6. For duplicates: Add 46 μl of the reaction mixture to a fresh microtube, and add 4 μl of each RNA dilution.
7. Flick the tube, and then spin down.
8. Pipette 25 μl of the reaction mixture with RNA into PCR tube strips, and spin down.
9. Set PCR program as follows:

   50 °C 30 min, 1 cycle

   95 °C 10 min, 1 cycle

   95 °C 15 s→60 °C 30 s→72 °C 30 s, 40 cycles

   72 °C 10 min, 1 cycle

3.5.2 Manipulating Cellular O-GlcNAcylation Levels

Inhibiting OGT Enzymatic Activity—Ac-5SGlcNAc

1. Seed 2.5×10^5 cells 24 h prior to treatment with Ac-5SGlcNAc.
2. Treat cells with control (DMSO) or 100 μM final concentration of Ac-5SGlcNAc in cell culture media for 48 h.
3. Collect lysates using appropriate methods, and perform Western blot for O-GlcNAc using CTD110.6- or RL2 O-GlcNAc-specific antibodies.

Inhibiting OGA Enzymatic Activity—6-Ac-Cas

1. Seed 2.5×10^5 cells 24 h prior to treatment with 6-Ac-Cas.
2. Treat cells with control or 100 μM final concentration of 6-Ac-Cas in cell culture media for 24 h.
3. Collect lysates using appropriate methods, and perform Western blot for O-GlcNAc using CTD110.6- or RL2 O-GlcNAc-specific antibodies.

3.5.3 Detecting O-GlcNAcylation on Specific Protein Substrates

1. Lyse cells with radioimmune precipitation assay buffer (150 mM NaCl, 1 % NP40, 0.5 % DOC, 50 mM Tris–HCl at pH 8, 0.1 % SDS, 10 % glycerol, 5 mM EDTA, 20 mM NaF, and 1 mM Na_3VO_4) supplemented with 1 μg/ml each of pepstatin, leupeptin, aprotinin, and 200 μg/ml PMSF.
2. Incubate lysates with antibody of protein predicted to be O-GlcNAcylated overnight at 4 °C.
3. Pre-clear Protein G Sepharose beads in 1 % BSA for 2 h at 4 °C on end-over-end rotator.
4. The next day, subject samples to immunoprecipitation using previously cleared Protein G Sepharose beads followed by 3× washes in 1 % PBS + Tween 20.
5. Resolve immunoprecipitated proteins by SDS-PAGE and transfer to a PVDF membrane.
6. Follow immunoblotting protocol as per the instructions above.
7. For positive control IP transcription factor Sp1 can be used since it is highly O-GlcNAcylated [22].
8. One can also repeat IP in cells treated with ±OGA or OGT inhibitor or OGT RNAi to ensure that the protein of interest is O-GlcNAcylated.
9. To potentially map specific O-GlcNAcylation sites on proteins please refer to other methods reviews [23] or contact mass spectrometry facilities that have expertise in this area (Medical University of South Carolina).

4 Notes

1. Plate cancer cell lines of interest, and treat around 50–70 % confluent as cells should be proliferating maximally at the time of glucose, lactate, and ATP assay.
2. Store 2-NBD glucose at −20 °C after reconstitution. Avoid light exposure of reconstituted reagent as well as treated cells to preserve fluorescent signal.
3. In addition to monitoring using flow cytometry, 2-NBDG can also be used to detect glucose uptake using confocal, high-resolution, or fluorescence microscopy.
4. Because lactate dehydrogenase degrades lactate, samples containing LDH (such as culture medium) should be kept at −80 °C for storage.
5. For lactate production assay, sample results may be normalized to standard curve or alternatively to total protein (μg) from protein lysate.

6. Follow the manufacturer's protocol for preparing ATP assay reagents if using kit for the first time, and once reconstituted, keep all reagents at −80 °C.
7. Samples need to be read in luminometer immediately following addition of ATP monitoring enzyme for best results.
8. For best general results during immunoblotting, dilute all primary antibodies in 3 % BSA in 1× TBS-T and 0.05 % sodium azide (as a preservative). For best results using secondary antibodies, dilute in 5 % powdered milk in 1× TBS-T.
9. For cell lines with low endogenous levels of HIF-1α, cells may have to be induced using hypoxia (1 % O2, 6 h) or pharmacological by treating cells with prolyl hydroxylase inhibitors (DMOG, $CoCl_2$) to be able to detect (immunoblot) or immunoprecipitate HIF-1α.
10. Following RNA extraction using RNeasy kit, store RNA at −80 °C for future use to prevent degradation.

Acknowledgements

We acknowledge Valerie L. Sodi for critical reading of this chapter. We thank previous members of Reginato lab for establishing some of these protocols. This work was supported in part by NIH-NCI grant RO1CA155413.

References

1. DeBerardinis RJ, Lum JJ, Hatzivassiliou G, Thompson CB (2008) The biology of cancer: metabolic reprogramming fuels cell growth and proliferation. Cell Metab 7:11–20
2. Vander Heiden MG, Cantley LC, Thompson CB (2009) Understanding the Warburg effect: the metabolic requirements of cell proliferation. Science 324:1029–1033
3. DeBerardinis RJ, Mancuso A, Daikhin E, Nissim I, Yudkoff M, Wehrli S, Thompson CB (2007) Beyond aerobic glycolysis: transformed cells can engage in glutamine metabolism that exceeds the requirement for protein and nucleotide synthesis. Proc Natl Acad Sci U S A 104:19345–19350
4. Wellen KE, Thompson CB (2010) Cellular metabolic stress: considering how cells respond to nutrient excess. Mol Cell 40:323–332
5. Hart GW, Housley MP, Slawson C (2007) Cycling of O-linked beta-N-acetylglucosamine on nucleocytoplasmic proteins. Nature 446:1017–1022
6. Caldwell SA, Jackson SR, Shahriari KS, Lynch TP, Sethi G, Walker S, Vosseller K, Reginato MJ (2010) Nutrient sensor O-GlcNAc transferase regulates breast cancer tumorigenesis through targeting of the oncogenic transcription factor FoxM1. Oncogene 29:2831–2842
7. Lynch TP, Ferrer CM, Jackson SR, Shahriari KS, Vosseller K, Reginato MJ (2012) Critical role of O-Linked beta-N-acetylglucosamine transferase in prostate cancer invasion, angiogenesis, and metastasis. J Biol Chem 287: 11070–11081
8. Yi W, Clark PM, Mason DE, Keenan MC, Hill C, Goddard WA 3rd, Peters EC, Driggers EM, Hsieh-Wilson LC (2012) Phosphofructokinase 1 glycosylation regulates cell growth and metabolism. Science 337:975–980
9. Itkonen HM, Minner S, Guldvik IJ, Sandmann MJ, Tsourlakis MC, Berge V, Svindland A, Schlomm T, Mills IG (2013) O-GlcNAc transferase integrates metabolic pathways to regulate the stability of c-MYC in human prostate cancer. Cancer Res 73(16):5277–5287
10. Shaw RJ, Cantley LC (2012) Decoding key nodes in the metabolism of cancer cells: sugar & spice and all things nice. F1000 Biol Rep 4:2

11. Zoncu R, Efeyan A, Sabatini DM (2011) mTOR: from growth signal integration to cancer, diabetes and ageing. Nat Rev Mol Cell Biol 12:21–35
12. Semenza GL (2011) Regulation of metabolism by hypoxia-inducible factor 1. Cold Spring Harb Symp Quant Biol 76:347–353
13. Whelan KA, Schwab LP, Karakashev SV, Franchetti L, Johannes GJ, Seagroves TN, Reginato MJ (2013) The oncogene HER2/neu (ERBB2) requires the hypoxia-inducible factor HIF-1 for mammary tumor growth and anoikis resistance. J Biol Chem 288: 15865–15877
14. Reginato MJ, Mills KR, Paulus JK, Lynch DK, Sgroi DC, Debnath J, Muthuswamy SK, Brugge JS (2003) Integrins and EGFR coordinately regulate the pro-apoptotic protein Bim to prevent anoikis. Nat Cell Biol 5:733–740
15. Tandon P, Gallo CA, Khatri S, Barger JF, Yepiskoposyan H, Plas DR (2011) Requirement for ribosomal protein S6 kinase 1 to mediate glycolysis and apoptosis resistance induced by Pten deficiency. Proc Natl Acad Sci U S A 108:2361–2365
16. Li L, Zhang L, Zhang X, Yan Q, Minamishima YA, Olumi AF, Mao M, Bartz S, Kaelin WG Jr (2007) Hypoxia-inducible factor linked to differential kidney cancer risk seen with type 2A and type 2B VHL mutations. Mol Cell Biol 27:5381–5392
17. Debnath J, Muthuswamy SK, Brugge JS (2003) Morphogenesis and oncogenesis of MCF-10A mammary epithelial acini grown in three-dimensional basement membrane cultures. Methods 30:256–268
18. Semenza GL (2011) Oxygen sensing, homeostasis, and disease. N Engl J Med 365: 537–547
19. Harris AL (2002) Hypoxia—a key regulatory factor in tumour growth. Nat Rev Cancer 2:38–47
20. Semenza GL (2010) HIF-1: upstream and downstream of cancer metabolism. Curr Opin Genet Dev 20:51–56
21. Whelan KA, Caldwell SA, Shahriari KS, Jackson SR, Franchetti LD, Johannes GJ, Reginato MJ (2010) Hypoxia suppression of Bim and Bmf blocks anoikis and luminal clearing during mammary morphogenesis. Mol Biol Cell 21:3829–3837
22. Yang X, Su K, Roos MD, Chang Q, Paterson AJ, Kudlow JE (2001) O-linkage of N-acetylglucosamine to Sp1 activation domain inhibits its transcriptional capability. Proc Natl Acad Sci U S A 98:6611–6616
23. Zachara NE (2009) Detecting the "O-GlcNAc-ome"; detection, purification, and analysis of O-GlcNAc modified proteins. Methods Mol Biol 534:251–279

Chapter 8

Targeted Genome Modification via Triple Helix Formation

Adele S. Ricciardi, Nicole A. McNeer, Kavitha K. Anandalingam, W. Mark Saltzman, and Peter M. Glazer

Abstract

Triplex-forming oligonucleotides (TFOs) are capable of coordinating genome modification in a targeted, site-specific manner, causing mutagenesis or even coordinating homologous recombination events. Here, we describe the use of TFOs such as peptide nucleic acids for targeted genome modification. We discuss this method and its applications and describe protocols for TFO design, delivery, and evaluation of activity in vitro and in vivo.

Key words Homologous recombination, Mutagenesis, Peptide nucleic acid (PNA), Triplex, Triplex-forming oligonucleotide (TFO), Site-specific gene editing

1 Introduction

1.1 Triplex-Forming Oligonucleotides

While double helices are key to the understanding and study of the biological sciences, nucleic acids are also capable of forming triple helices. In fact, before the establishment of the double-helical nature of DNA, Linus Pauling proposed a triple-helix structure [1]. Felsenfeld et al. demonstrated the possibility of triple helix formation when they noted that polyU and polyA RNA strands could bind in a 2:1 ratio [2]. Triplex-forming oligonucleotides, or TFOs, can form similar triple helices. TFOs can bind in the major groove of duplex DNA in a polypurine/polypyrimidine run, with reverse Hoogsteen hydrogen bonds antiparallel to a polypurine strand of a DNA duplex or with Hoogsteen bonds in a parallel orientation to the purine strand (Fig. 1a).

While both DNA and RNA can form triple-helix structures, novel synthetic nucleic acid analogues can also be used for TFO formation (Fig. 1b). Peptide nucleic acids (PNAs) are synthetic compounds with a neutral polyamide rather than charged phosphodiester backbone. They are more resistant to protease and nuclease degradation and can bind more tightly to DNA and RNA [3, 4]. Gamma PNA molecules, which feature a pre-organized

Narendra Wajapeyee (ed.), *Cancer Genomics and Proteomics: Methods and Protocols*, Methods in Molecular Biology, vol. 1176, DOI 10.1007/978-1-4939-0992-6_8, © Springer Science+Business Media New York 2014

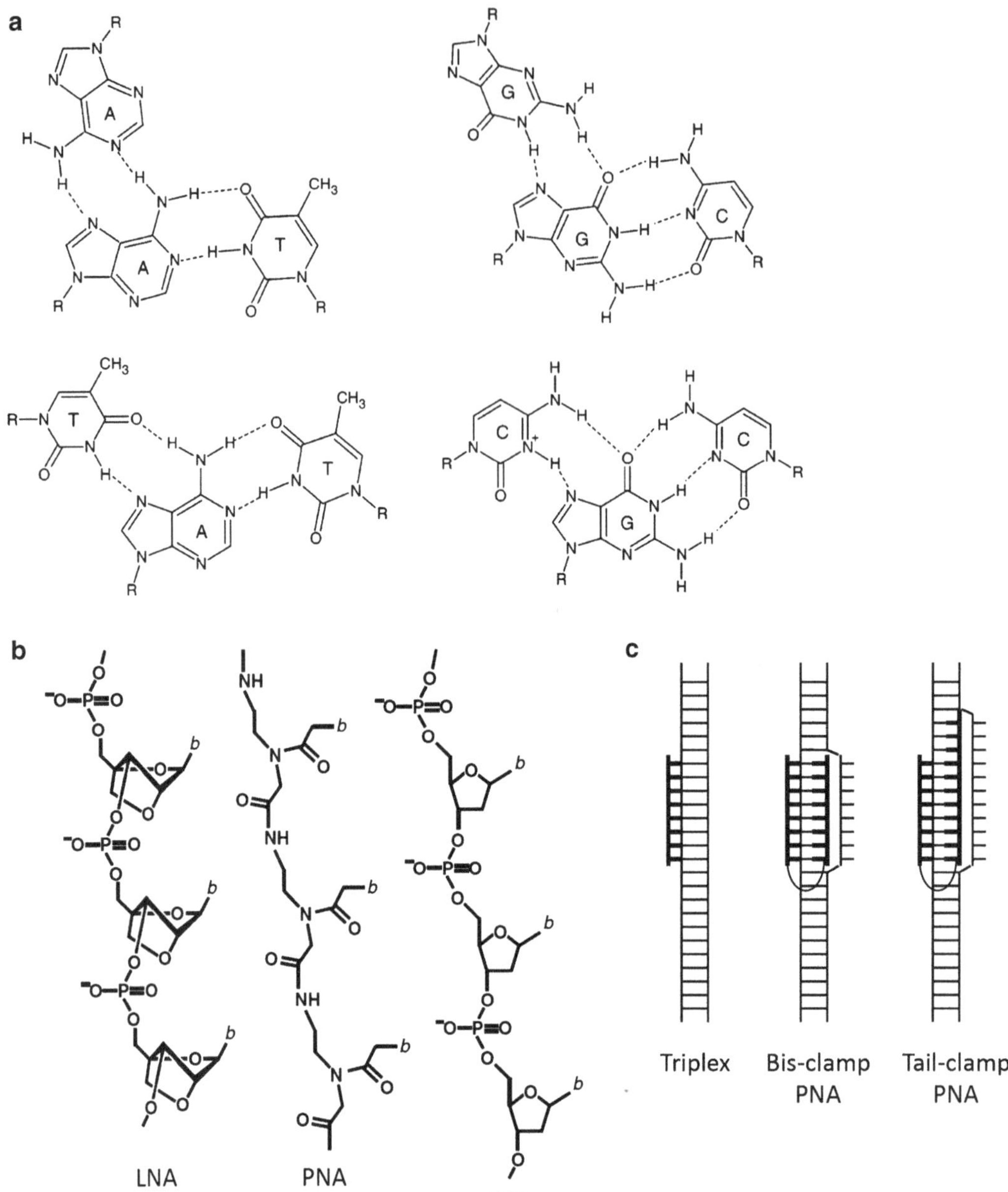

Fig. 1 Chemical structures. (**a**) Example of Hoogsteen bond formation in triple helix, purine (AAT and GGC) and pyrimidine (TAT and CGC) motif. From Gorman et al. (2001) *Current Molecular Medicine* reprinted in Schleifman et al. Methods in Molecular Biol Vol. 435. (Schleifman, Chin et al. [59]). (**b**) Examples of novel synthetic nucleic acid analogues. From Briones et al. Anal Bioanal Chem 2012. (**c**) Examples of triplex structures

conformation, have increased binding to target DNA [5, 6]. Mini-PEG or other modifications may also increase PNA binding to target DNA [7]. Locked nucleic acids (LNAs) are another synthetic oligonucleotide, with a bridge between the 2′ oxygen and

4′ carbon, "locking" the ribose in a fixed 3′-endo configuration [8–10]. This rigid conformation reduces barriers to binding by lowering the binding entropy [11].

PNAs and other molecules are capable of forming unique structures with DNA (Fig. 1c). These structures include triplexes consisting of PNA binding with DNA through Hoogsteen base pairs at homopurine/pyrimidine stretches. In addition, two PNA molecules connected by a flexible linker can form a bis-PNA "clamp" that can form a PNA/DNA/PNA triplex at the target site [12]. Tail clamp PNAs (tcPNAs) have an extended 5–10 bp "tail" that forms a PNA/DNA duplex in addition to a PNA/DNA/PNA "clamp," allowing for more specific binding without the need for a long (15–17 mer) homopurine/pyrimidine stretch [13, 14]. Novel duplex structures are also possible, such as PNA strand-invasion duplexes and pseudocomplementary PNAs (pcPNAs) capable of double-duplex invasion [15]. Strand invasion and cellular uptake can be enhanced by addition of positively charged lysine residues [16], and cell-penetrating peptides (CPPs) can be added to molecules to enhance the uptake [17, 18]. The base substitute pseudoisocytosine (J) can be used in place of cytosine to encourage Hoogsteen bond formation in a pH-independent fashion [12]. These unique synthetic molecules have numerous applications, some of which we describe below.

1.2 Applications of TFOs

TFO binding has been shown to inhibit transcription, replication, and protein binding to DNA [19–21]. In addition, TFOs tethered to mutagens can promote DNA damage in a sequence-specific fashion and induce mutagenesis [22–25]. More recently, researchers have demonstrated that TFOs can mediate site-specific gene modification, both in vitro and in vivo [26–29]. TFOs can also be used for splice site correction, for example in Duchenne muscular dystrophy [30] or beta-thalassemia [31]. Triplex-forming molecules have also recently been used for the suppression of oncogenes and proto-oncogenes to reduce cancer cell growth. For example, TFOs have been used to decrease MET expression and induce cell death in hepatoma cells [32], reduce cell proliferation by binding to Ki-ras [33] and bcl-2 [34–36], and HER-2/neu [37].

In this chapter we focus on the use of triplex-forming molecules to mediate gene modification and novel methods for TFO delivery that can be used for transfer of diverse nucleic acids. While introduction of an oligonucleotide homologous to a target gene may lead to recombination at low levels, use of TFOs can enhance recombination frequencies, leading to targeted, specific editing of endogenous human genes. PNA TFOs have recently been used to mediate site-specific gene editing in a beta-thalassemia-associated site, leading to heritable modification in primary human hematopoietic stem cells [26, 38, 39]. tcPNA molecules have been used to

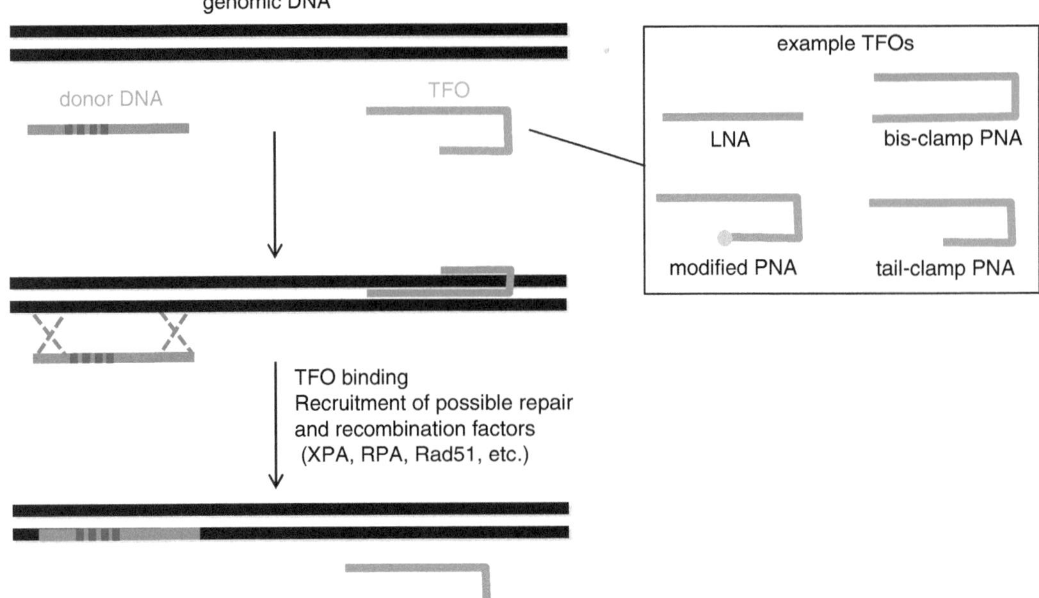

Fig. 2 Schematic of genome modification using triplex-forming oligonucleotides. Triplex formation by site-specific intracellular TFO binding recruits factors that can induce recombination of a single-stranded donor DNA, encoding a desired modification, at a nearby genomic location

modify the human *CCR5* gene, leading to the production of cell lines resistant to HIV-1 infection [27]. Conjugation of PNA molecules to CPPs has allowed for direct in vivo targeting of genes for site-specific mutagenesis in hematopoietic stem cells [28], and encapsulation of PNA molecules in biodegradable polymer nanoparticles has allowed for editing of primary human cells with lower toxicity and direct in vivo gene editing [29, 40]. This type of directed gene modification could also be used for cancer gene therapy or creation of selectively edited cell lines for the study of cancer biology.

1.3 Mechanism of Triplex-Induced Genome Modification

PNA TFOs form triplexes by binding with high affinity and specificity in the major groove of a complementary strand of duplex DNA. The stable, abnormal PNA/DNA/PNA triple helix is recognized by a cell's own DNA repair machinery, sensitizing the surrounding DNA for homologous recombination [38]. PNAs exhibit stable, high-affinity binding to DNA and genetic specificity [41], making them advantageous for creating heritable changes in targeted genes. Intracellular delivery of a site-specific PNA has also been shown to induce recombination of a short, single-stranded donor DNA molecule, encoding a desired modification, with a nearby genomic site, shown schematically in Fig. 2 [38].

Mechanisms of TFO-induced DNA repair and recombination have been previously reviewed [42, 43]. Multiple investigations indicate that the nucleotide excision repair (NER) pathway has a role in recognizing and repairing triplex structures [25, 26, 38, 44]. Evidence suggests that the NER factors xeroderma pigmentosum group A (XPA) and replication protein A (RPA) bind specifically as a complex to cross-linked triplex structures [44]. XPA deficiencies have also been shown to decrease TFO-induced plasmid recombination [45], decrease PNA-induced plasmid repair synthesis, and decrease PNA-induced plasmid recombination with an ssDNA donor [38]. While there appears to be a clear association between the NER pathway and TFO-induced repair and recombination, some evidence suggests that other repair pathways may be involved in metabolizing triplex structures [45]. Additionally, it was found that overexpression of Rad51, a homologous recombination factor, increased TFO-induced recombination of tethered donor DNA into a shuttle vector [46]. Furthermore, Rad51 deficiency resulted in no TFO-induced recombination of tethered donor DNA [46].

In addition to being an important tool for instigating DNA repair, triplex-forming molecules may also prove to be an important tool to study DNA repair machinery at different target sites. Current evidence suggests that the repair of TFO-associated lesions is dependent on the NER pathway, while other repair pathways such as homologous recombination and transcription-coupled repair may also be involved in TFO-induced recombination. Although the mechanisms by which PNA TFOs are able to augment recombination frequencies of donor DNA fragments into genomic targets are not clearly defined, this is an ongoing area of research. Further studies aimed at understanding how triplex structures are recognized and repaired will be instrumental in improving the efficacy of targeted genome modification of disease-causing genes.

1.4 Overview of Techniques

In this chapter, we discuss the use of triplex-forming oligonucleotides and novel delivery vehicles for site-specific genome editing. We describe the selection of target sites for gene editing, evaluation of binding affinity of TFOs, creation of delivery tools for oligonucleotide transfer into target cells, and evaluation of mutagenesis and site-specific genome editing.

2 Materials

1. Oligonucleotides: Oligonucleotides discussed here can be ordered from Midland Certified Reagent Company Inc. (Midland, TX) or other vendors. Single-stranded donors should be protected by three phosphorothioate internucleoside linkages at both the 5′- and 3′-end to prevent degradation and should be purified by reversed-phase high-performance liquid chromatography.

2. PNAs: PNAs can be ordered from Bio-Synthesis (Lewisville, TX) or Panagene (Daejeon, South Korea).
3. Gel shift assays: Taq polymerase—Invitrogen (Carlsbad, CA). Qiagen Gel extraction kit and QIAquick PCR purification kit (Venlo, Limburg), T4 DNA ligase, Tris–EDTA buffer—pH 8. 10 mM Tris, bring to pH 8.0 with HCl and 1 mM EDTA, and polyacrylamide gel—40 % 19:1 bis:acrylamide, TBE buffer (1 L of 5× stock—54 g Tris base, 27.5 g boric acid, 20 mL of 0.5 M EDTA pH 8.0), EDTA, 10 % ammonium persulfate, TEMED.
4. Silver staining: Ag solution—0.1 % silver nitrate in dH_2O, developer solution—3 % potassium carbonate plus 250 μL formalin and 125 μL 10 % sodium thiosulfate per liter.
5. Nucleofection: Amaxa Nucleofection Kits—Lonza Group (Basel, Switzerland), Geneporter 2—Gene Therapy Systems (San Diego, CA), or BTX Electro Square Porator—BTX (Holliston, MA). StemSpan culture media—STEMCELL Technologies Inc. (Vancouver, Canada), RPMI media, FBS, and L-glutamine.
6. Peptide–PNA conjugates: Penetratin 1—Qbiogene, 0.1 M DTT, NAP5 filtration column—GE Healthcare Life Sciences (Piscataway, NJ).
7. PLGA: 50:50 ester-terminated PLGA, 0.95–1.2 g/dl, can be ordered from LACTEL absorbable polymers—DURECT corporation (Birmingham, AL).
8. Reporter systems: DMEM—Life Technologies (Carlsbad, CA), 10 % FCS, phenol, chloroform, isoamyl, Qiagen Gel Extraction Kit (Venlo, Limburg), RPMI media—Sigma Aldrich (St. Louis, MO), Ficoll-Paque—GE Healthcare Life Sciences (Piscataway, NJ).
9. Mouse model: EGFP-654 transgenic mice from the laboratory of Ryszard Kole (Chapel Hill, NC) [31].

3 Methods

3.1 TFO Design and Synthesis

3.1.1 Selection of Target Site and TFOs

1. As noted above, Hoogsteen bonding for triplex formation requires homopurine/pyrimidine stretches. These should be identified near the site targeted for genome modification within a few hundred base pairs of the target [47].
2. TFO-binding sites should be 13–30 bp and PNA-binding sites should be 8–10 bp, with 5–10 bp clamp for added specificity and increased binding affinity if desired [27]. TFOs will bind in the parallel orientation to the purine-rich strand of the target or in the antiparallel orientation with the polypurine strand.

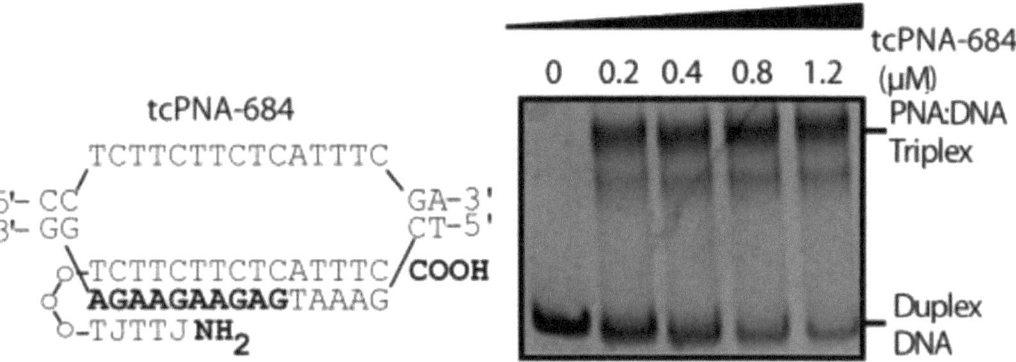

Fig. 3 Example of target site and gel shift assay. Sample sequence and tail clamp PNA binding to a site in the cystic fibrosis transmembrane receptor gene. Reprinted from Schleifman et al. (Schleifman, Bindra et al. [27])

3. Several modifications are also available. Psoralen can be used to induce cross-links [48] and can be conjugated to the TFO with phosphoramidates. bis-PNA clamps consist of two PNAs connected by a linker such as 8-amino-3,6-dioxaoctanoic acid (O). Lysine residues can be added to either or both ends to enhance the uptake and binding, and pseudoisocytosine (J) can be used to reduce pH dependence of cytosine N3 protonation. An example of a bis-PNA tail clamp with terminal lysines is given in Fig. 3.

3.1.2 Choice of Donor DNA

1. Single-stranded DNA donors should be homologous to the site desired for modification, except for the desired base pair change(s). Antisense or sense donors can be designed and can be between 30 and 100 nucleotides in length. Six base pair changes are preferred to easily detect the change by allele-specific PCR (AS-PCR).
2. First and last three bases should use phosphorothioate linkages to inhibit exonuclease degradation.

3.1.3 Oligonucleotides

1. PNA may be purchased or synthesized using Boc-protected monomers on solid support as described by Christensen et al. [49]. Molecules can be purified by reverse-phase HPLC and characterized by MALDI-TOF.

3.2 Evaluation of Binding with Gel Shift Assays

A gel shift assay can be used to confirm the binding of PNA molecules to their desired target sequences of DNA. To perform this assay, plasmids that contain a 150–200 base pair section of the target gene including one or two of the potential PNA target sequences should be created. Select the 150–200 base pair region of interest, and select primers to amplify this region. Amplify using Taq DNA polymerase as per the manufacturer's protocol. Run PCR products on 1.1 % agarose gel. Extract PCR products using

Qiagen Gel Extraction Kit as per the manufacturer's protocol. Identify restriction enzyme sites around the region of interest that are also present in a plasmid of your choice (3–4 kb). Digest both the PCR products and the plasmids separately overnight. Purify the PCR products and digested plasmid using the QIAquick PCR Purification Kit as per the manufacturer's protocol. Ligate the PCR products and plasmid using T4 DNA ligase (or other ligases of choice). Submit plasmids for sequencing to verify insert.

3.2.1 Binding Reaction

To perform the binding reaction and allow the PNA to bind to its target sequence, combine various amounts of PNA diluted to 10 μM for final PNA concentrations of 0, 0.2, 0.4, 0.8, and 1.2 μM, 5 μL of 400 ng/μL of plasmid DNA, 6.7 μL of 30 mM of KCl for a final concentration of 10 mM, and enough TE (pH 8) to bring the final reaction volume to 20 μL. These reactions should be left to incubate overnight at 37 °C.

3.2.2 Digest

Once the PNA has bound its target on the plasmid, a 150–200 base pair fragment of the target gene is cut out from the plasmid so that it can be run on a gel (*see* **Note 1**). To perform this digestion reaction, the 20 μL of the binding reaction should be combined with 1 μL of each enzyme used to insert the target gene into the plasmid, 3 μL of the appropriate buffer for these enzymes, 0.3 μL of 100× BSA, and 4.7 μL of dH_2O to bring the final reaction volume to 30 μL. The digestion reaction should be incubated at 37 °C for 2 h.

3.2.3 Running the Gel

To detect binding, digested DNA must be run on a polyacrylamide gel, which allows DNA fragments to be separated based on size and charge. To make the 8 % native gel, mix 10 mL of 40 % 19:1 bis:acrylamide, 10 mL 5× TBE buffer, 200 μL 0.5 M EDTA, 29.25 mL dH_2O, 500 μL 10 % ammonium persulfate (APS), and 50 μL TEMED for a final volume of 50 mL. Using a large pipette, transfer this mixture to an assembled gel apparatus. Allow approximately 30 min for the gel to solidify, and add TBE to flush the wells. To 15 μL of each digest reaction, add 3 μL 6× DNA loading dye, and load the 18 μL onto the gel, attempting to keep the samples as close to the middle of the gel as possible, as the sides of the gel may become distorted. Load 100 and 10 bp DNA ladders onto the gel on either end of the samples. Run the gel at 33 mA for 30 min to allow the loaded DNA to settle into the wells and then run at 20 mA for approximately 4 h to ensure good separation.

3.2.4 Visualizing the Gel

Once the gel has run, it must be stained to visualize the DNA fragments present. When disassembling the gel apparatus, leave the gel on one glass plate for support and immerse this in fresh Ag solution (0.1 % silver nitrate in dH_2O) for 10 min, making sure not to agitate it. Pour off the Ag solution into a bottle for future use (can be used up to six times). Rinse the gel with dH_2O and develop for 1 min with freshly made developer solution (3 % potassium carbonate plus 250 μL formalin and 125 μL 10 % sodium thiosulfate per liter). If bands cannot be seen, the gel can be left to develop for more time, though this does not usually take more than 15 min. Once developed, developer solution should be poured out and the gel can be "picked up" with Whatman filter paper by laying the paper on the gel and allowing the gel to adhere to it. The gel can then be covered in Saran wrap and dried so that it can be preserved.

When examining the stained gel, one, two, or several bands may be observed. A lower band indicates unbound DNA, whereas successful binding is demonstrated by a shift of the DNA band upward on the gel, since the charged PNA molecule slows down the transit of the complex. Therefore, the absence of a lower band indicates that all the DNA has been bound by PNA and reflects optimal binding. Several bands may be present due the fact that PNA can infiltrate the DNA molecule to bind in different configurations, each of which migrates differently through the gel.

3.3 Delivery Methods

Several options exist for delivery of TFOs to cells, including nucleofection, use of CPPs, and delivery in biodegradable nanoparticles. Nucleofection can be accomplished using commercially available kits, but can only be used in vitro and may lead to high levels of cell death compared to nanoparticle delivery or other methods [40]. Conjugation of TFOs to CPPs or delivery in biodegradable nanoparticles allows for transfection with lower cell death and direct in vivo nucleic acid delivery [18, 28, 29, 40].

3.3.1 Nucleofection

1. Nucleofection can be used for delivery of TFOs to certain cell types, and optimal procedures will depend on cell type and molecules to be delivered. Amaxa Nucleofection Kits can be used for nucleofection of primary human hematopoietic cells with PNA and DNA [26, 27]. Amaxa nucleofection may be superior to some other reagents in several different cell systems [50]. Geneporter 2 or BTX Electro Square Porator can also be used according to the manufacturer's instructions for delivery of TFOs.
2. For primary human CD34+ cells, 1×10^6 cells can be nucleofected in 100 μL complete media with 0.2 nmol DNA + 0.8 nmol PNA using the Amaxa Human CD34+ Nucleofection Kit as per the manufacturer's protocol. Cells can then be resuspended in StemSpan culture media with cytokines for expansion.
3. THP1 and K562 cells can be electroporated with TFO in 100 μL PBS using a BTX Electro Square Porator ECM 830 at 350 V, 12 ms, and 1 pulse. Cells can be resuspended in RPMI media with 10 % FBS and L-glutamine for expansion.

3.3.2 Peptide–PNA conjugates

1. Conjugation of TFOs to CPPs may greatly enhance the uptake—we present one strategy here. PNAs can be conjugated to a CPP such as Antp (Penetratin 1) via a reducible disulfide linkage, with activation of the peptide with a pyridyl disulfide at the N-terminus. A 3′-thiol modified TFO [18, 28].
2. Dissolve a 3′-thiol-modified TFO in 0.1 M DTT and leave overnight at 37 °C. Apply the solution to a NAP5 filtration column equilibrated with dH_2O and elute into a siliconized Eppendorf with activated Antp. Add methanol dropwise if precipitation is observed, and vortex. Heat the reaction at 65 °C for 15 min, incubate for an additional hour at 37 °C, and evaporate methanol if needed.

3.3.3 Nanoparticles

The use of an engineered biodegradable polymer delivery system for site-specific genome editing was initially described in the literature in 2011 when McNeer et al. developed methods to formulate poly(lactic-co-glycolic acid) (PLGA) nanoparticles containing PNA and DNA to be used for enhanced delivery to hematopoietic stem and progenitor cells [40]. PLGA is an FDA-approved biocompatible and biodegradable polymer used clinically for delivery of drugs, including drugs for treatment of prostate cancer (Lupron Depot and Trelstar). Previous work has shown that PLGA nanoparticles can provide intracellular delivery of nucleic acids and oligomers, including plasmid DNA for transfection of cell lines [51, 52] and siRNAs for sustained gene silencing [53]. McNeer et al. first demonstrated that PLGA nanoparticles could be used as an intracellular delivery system for triplex-forming PNAs and short, donor DNAs in primary human CD34+ cells. They showed that the PNA/DNA PLGA nanoparticle delivery system was efficient and nontoxic but also could be used for specific PNA-mediated recombination in the human β-globin locus. In addition, they demonstrated modification in the human *CCR5* gene using a *CCR5*-targeted PNA with a *CCR5* donor DNA [40]. Site-specific genome editing of the *CCR5* and β-globin genes in human hematopoietic cells via triplex-forming PNA and donor DNA nanoparticles has also been described in vivo in chimeric mice [29]. PLGA nanoparticles containing an anti-microRNA-155 (anti-miR-155) PNA have additionally been shown to slow the growth of miR-155 "addicted" tumors in vivo in a mouse model of lymphoma [54].

Nanoparticles for these studies were formulated using a previously described [55] double-emulsion solvent evaporation technique that was modified to allow for the encapsulation of PNA alone or PNA and DNA [40]. General methods, as well as an example of specific instructions for making a 40 mg batch of PLGA particles, are described below (*see* **Note 2**).

1. Dissolve 50:50 ester-terminated PLGA, 0.95–1.2 g/dl, in an organic solvent such as dichloromethane or another organic solvent partially miscible in water (e.g., ethyl acetate, benzyl

alcohol, propylene carbonate) (1 mg polymer/10 μL organic phase). For example, dissolve 40 mg of PLGA in 400 μL of dichloromethane overnight. Note: A different polymer blend containing PLGA and poly(beta-amino)ester (PBAE) can also be used as a PNA/DNA delivery vehicle [56].

2. Add the encapsulant, PNA/DNA in dH_2O, dropwise to the polymer solution while stirring. For example, make a mixture of 10 μL of donor DNA in a 2 mM solution with 10 μL of PNA in a 1 mM solution with 10.8 μL water. This 30.8 μL solution would be added to the 400 μL polymer solution above (*see* **Note 2**).
3. Probe sonicate the first emulsion for 30 s at 38 % amplitude to further emulsify. Amplitude and frequency settings should be optimized for the sonicator probe used.
4. Add the first emulsion dropwise to 5 % polyvinyl alcohol (use twice the initial volume of organic solvent) while stirring. Any desired surface modifiers (e.g., DSPE-PEG, avidin) should be dissolved in 5 % polyvinyl alcohol (second emulsion).
5. Probe sonicate the second emulsion for an additional 30 s.
6. Transfer the final mixture to a stirring beaker of 0.3 % polyvinyl alcohol. For a 40 mg batch, use 20 mL of 0.3 % PVA. Stir at room temperature for 3 h to evaporate the dichloromethane.
7. Collect the nanoparticles by spinning for 10 min at 16,000×*g* at 4 °C.
8. Wash the nanoparticles in cold dH_2O. The number of washes can vary depending on the intended use of the nanoparticles. Three washes are normally sufficient for cell culture and animal studies. Mix the nanoparticles with 10 mL of cold dH_2O and spin at 16,000×*g* for 10 min for each wash.
9. After washing the nanoparticles, resuspend in dH_2O, and freeze at −80 °C for at least 1 h.
10. Lyophilize the nanoparticles for 2–3 days.
11. Morphology of the nanoparticles can be analyzed using an XL-30 scanning electron microscope (FEI, Hillsboro, Oregon) or comparable instrument. To image the particles, smear them on an imaging stub, and sputter-coat with palladium or gold. ImageJ software analysis can be used to determine particle diameter. The hydrodynamic diameter of particles in aqueous solution can also be determined by light scattering using instruments such as the Malvern Zetasizer (*see* **Note 3**).

3.4 Evaluation of Mutagenesis

3.4.1 Reporter Systems

A mouse cell line (AV16) with a multiple chromosomally integrated λ supFG1 shuttle vector carrying the *supFG1*mutation reporter gene can be used to assess mutagenesis mediated by TFO AG30 (AGG AAG GGG GGG GTG GTG GGG GAG GGG GAG) or variations [18, 57, 58]. We have also described the use of a luciferase-based assay in a previous *Methods* chapter [59] Subheading 3.6. Evaluation of Induced Recombination: Luciferase Assay.

1. Cells are grown in DMEM with 10 % FCS with G418 at 0.8 mg/mL.
2. Add TFO to the cells, and collect genomic DNA after allotting time for mutation induction (normally 2–4 days posttreatment).
3. Prepare high-molecular-weight DNA by lysis with 10 mM Tris (pH 8), 100 mM EDTA, 0.1 % SDS, and 50 μg/mL proteinase for 3 h at 37 °C, followed by phenol extraction (25 phenol:24 chloroform:1 isoamyl) and ethanol precipitation.
4. Rescue phage vector DNA into phage particles using in vitro packaging extracts made from restriction deficient lysogen (NM759). Grow the phage on *E. coli* containing an amber mutation in the *lacZ* gene with 5-bromo-4-chloro-3-indolyl-β-D-galactopyranoside and isopropyl-β-D-thiogalactopyranoside.
5. Phage with functional *supF* genes will produce blue plaques, but phage with mutations will produce colorless plaques. The ratio of colorless to total plaques gives an estimation of mutation frequency.

3.4.2 Sequencing Strategies

1. Deep sequencing may be used for analysis of mutagenesis in a gene of choice.
2. Bar-coded primers, with a 6 bp unique barcode per sample, should be used to amplify a 100 bp region of interest using high-fidelity platinum taq polymerase with provided buffers (*see* **Note 4**). Run the PCR products on a 1 % agarose gel, and extract using the Qiagen gel extraction kit. Samples with different barcodes may be pooled for sequencing.
3. Samples can be ligated to adapters and sequenced on Illumina HiSeq platform using 75 base pair paired-end reads. Preparation for sequencing will vary depending on the sequencing platform used.

3.5 Evaluation of Recombination

3.5.1 Reporter Systems

A mouse reporter system developed by Roberts et al. [31] that ubiquitously expresses a modified eGFP pre-mRNA containing an aberrantly spliced β-globin intron (IVS2-654) can be used to evaluate PNA/DNA-mediated recombination in vivo. Nanoparticles containing PNA and DNA directed at the incorrect 5′ splice site can be delivered to the EGFP-654 transgenic reporter mice in an attempt to correct the aberrant splice site via site-specific gene modification. Correction of the aberrant splice site would result in the expression of eGFP in a cell that was correctly modified. This mouse model has previously been used to demonstrate splice switching using LNA molecules to bind at the aberrant splice site [31].

1. Design desired PNA and DNA molecules (described in Subheadings 3.1.1 and 3.1.2, respectively) to target the aberrantly spliced β-globin intron (IVS2-654). The DNA should contain the sequence for the correct splice site. We have recently used the following molecules [29]: donor DNA, 5′-AAAGAATAACAGTGATAATTTCTGGGTTAAGG CAATAGCAATATCTCTGCATATAAATAT-3′; 654PNA1, N terminus-KKK-JTTTJTTTJTJT-OOO-TCTCTT TCTTTCAGGGCA-KKK-C terminus; and 654PNA2, N terminus-KKK-JJJTJJTTJT-OOO-TCTTCCTCCCA-CAGCTCC-KKK-C terminus.
2. Formulate and image nanoparticles as described above in Subheading 3.3.3 and deliver to mice via preferred delivery route (e.g., tail vein injection, retro-orbital injection, intraperitoneal, intranasal).
3. After delivering desired dose of nanoparticles, sacrifice the mice and harvest organs or cells of interest.
4. Mechanically disrupt the organs in RPMI-1640 media, and strain the cells through a cell strainer into a 50 mL conical tube.
5. Next, separate any red blood cells from the harvested organs by Ficoll-Paque separation (2:1 ratio of cells to Ficoll-Paque). Slowly layer the cells onto the Ficoll-Paque in a 50 mL conical tube. Spin at 1,000 × *g* for 20 min. Collect the mononuclear cell layer and wash with dPBS. Collect the cell pellet, and fix desired amount of cells by resuspending in 4 % paraformaldehyde.
6. Look for the expression of eGFP (which indicates correction of aberrant splice site by recombination) by using fluorescent-associated cell sorting (FACS).
7. Harvested cells can also be co-stained using fluorescently labeled antibodies directed against a cell population of interest. Any antibodies used for co-staining should be detected in a channel other than the channel used to detect eGFP fluorescence.
8. Recombination can also be detected by using an allele-specific mRNA PCR technique, described below in Subheading 3.5.2.
9. Recombination can additionally be evaluated by imaging cells or tissue sections for expression of the eGFP modification using a fluorescence microscope.

3.5.2 Allele-Specific PCR

To detect different DNA sequences corresponding to the original and corrected gene, an allele-specific PCR method can be used.

1. Primer design—When selecting a set of two primers for allele-specific PCR, the same gene-specific reverse primer should be used, though the forward primers should be designed such

that their 3′ ends differ and correspond to either the original or the corrected gene sequence. Thus, under the correct conditions, each primer will only be able to amplify its corresponding sequence of DNA. Primers are usually about 20–30 base pairs in length, with roughly equal salt-adjusted melting temperatures between 60 and 70 °C. Increasing the G/C content of the primers raises the melting temperature, since there are stronger hydrogen-bonding interactions between these bases. Alternately, the same gene-specific forward primer may be used with an allele-specific reverse primer. Forward and reverse primers should be selected with a span of hundreds of base pairs between them.

2. PCR testing—To determine the optimal AS-PCR conditions, a gradient technique can be used to select the correct annealing temperature that will allow for each primer set to only amplify its corresponding DNA sequence (*see* **Note 5**). To test PCR conditions, plasmids should be designed to contain a segment of DNA corresponding to the entire section of DNA to be amplified, with either the original or the corrected DNA sequence present. It may be necessary to use site-directed mutagenesis to create one sequence from the other. Using the gradient technique four reactions can be set up; two primers sets can be paired with either the matching or the mismatched plasmid. Then these four reactions can be run simultaneously using a variety of annealing temperatures, usually from about 10 °C below to 10 °C above the primer melt temperature. Running the PCR products on an agarose gel and staining DNA with ethidium bromide for visualization under UV light can allow for the best conditions to be selected. This technique can lead to the detection of DNA sequences at plasmid concentrations as low as 0.002 ng/μL.
3. mRNA PCR—To modify this technique so that it can be used to detect specific mRNA sequences, the same allele-specific forward primers can be used, but the gene-specific reverse primer should be selected such that it corresponds to an exon that is separated from the location of the forward primer by an intron. This will ensure that only the mRNA sequence is short enough to be amplified by the primers for a given elongation time, with the result that the amplified product will only correspond to mRNA and not DNA.
4. qRT-PCR—To convert this technique into a real-time method to analyze levels of gene correction in a more quantitative fashion, fluorescent dyes that bind to DNA can be added to the PCR reaction mixture. Using a thermal cycle with the ability to read levels of fluorescence emitted by dyes such as SYBR Green, which is used to monitor DNA accumulation, and

ROX, used as a reference dye, one can plot the relative amounts of DNA amplified over time.

5. When attempting to optimize allele-specific PCR conditions, many problems may be encountered. A lack of PCR products may point to annealing temperatures that are too high, whereas nonspecific products indicate an annealing temperature that is too low. Another important factor that can be manipulated is Mg^{2+} concentrations, since this is an important cofactor for the DNA polymerase enzyme. To circumvent problems, a touchdown technique can be utilized, in which the annealing temperature is set to a high temperature during the PCR cycle but then decreased every cycle until landing on a lower temperature that is used for the remainder of the cycles. This helps ensure that the first PCR products amplified, and thus those that are preferentially amplified during future cycles, have the highest specificity.

3.5.3 Sequencing Strategies

1. Cells may be cloned using a limiting dilution strategy to identify clones with the desired modification using AS-PCR. Positive clones can then be harvested for genomic DNA isolation and PCR of a 100–200 bp region surrounding the modification site and amplicons submitted for regular sequencing.
2. Deep sequencing can be used for analysis of modification in the target gene and off-target sites. Primers should flank a 100 bp region with the targeted modification in the center. As described above, bar-coded amplicons can be sequenced on the Illumina HiSeq platform (*see* **Note 5**).

4 Notes

1. Gel shift assay: A 90–100 base pair segment for binding may also be created by PCR for binding assay, rather than use of plasmid. This may change optimal binding conditions, however.
2. Use of nanoparticles: Different cell types may require different dosages of nanoparticles, so an initial experiment should include a dose–response curve, including cell survival measures. Fluorescent dyes such as Coumarin 6 in nanoparticles (dissolved in the DCM phase of creating particles) can be used to track the uptake of particles in cells by FACS or confocal imaging (Fig. 4).
3. Nanoparticle formulation: SEM imaging and controlled-release profiles will help determine whether nanoparticles have appropriate morphology and loading.
4. Deep sequencing: Double bar-coding will allow the researcher to discard chimeric reads from analysis. In addition, it is

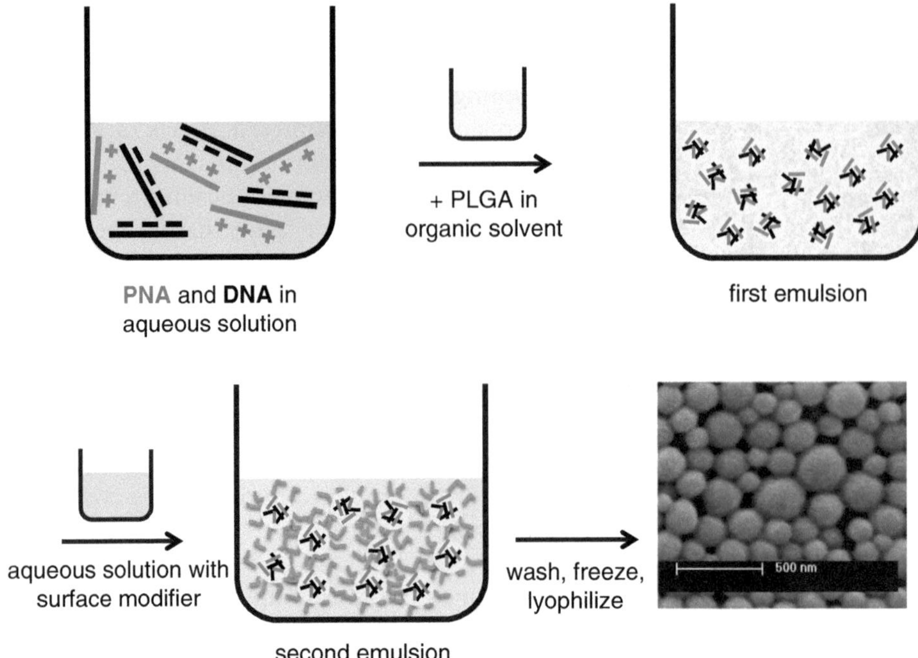

Fig. 4 Schematic of nanoparticle formulation. PNA and DNA were loaded into nanoparticles using a previously described double-emulsion solvent evaporation technique (Fahmy, Stamstein et al. [55]; McNeer, Chin et al. [40]). A scanning electron micrograph shows a sample batch of PNA/DNA-loaded PLGA nanoparticles

important to examine the background level of mutation in non-treated samples.

5. AS-PCR: It may be necessary to attempt multiple primer concentrations and PCR conditions to optimize the protocol.

References

1. Pauling L, Corey RB (1953) A proposed structure for the nucleic acids. Proc Natl Acad Sci U S A 39(2):84–97
2. Felsenfeld G, Rich A (1957) Studies on the formation of two- and three-stranded polyribonucleotides. Biochim Biophys Acta 26(3): 457–468
3. Nielsen PE, Egholm M, Buchardt O (1994) Peptide nucleic acid (PNA). A DNA mimic with a peptide backbone. Bioconjug Chem 5(1):3–7
4. Nielsen PE (1999) Peptide nucleic acid. A molecule with two identities. Acc Chem Res 32(7):624–630
5. He G et al (2009) Strand invasion of extended, mixed-sequence B-DNA by gammaPNAs. J Am Chem Soc 131(34):12088–12090
6. Rapireddy S, Bahal R, Ly DH (2011) Strand invasion of mixed-sequence, double-helical B-DNA by gamma-peptide nucleic acids containing G-clamp nucleobases under physiological conditions. Biochemistry 50(19):3913–3918
7. Bahal R et al (2012) Sequence-unrestricted, Watson-Crick recognition of double helical B-DNA by (R)-miniPEG-gammaPNAs. Chembiochem 13(1):56–60
8. Kumar R et al (1998) The first analogues of LNA (locked nucleic acids): phosphorothioate-LNA and 2'-thio-LNA. Bioorg Med Chem Lett 8(16):2219–2222
9. Koshkin AA et al (1998) LNA (Locked Nucleic Acids): synthesis of the adenine, cytosine, guanine, 5-methylcytosine, thymine and uracil bicyclonucleoside monomers, oligomerisation,

and unprecedented nucleic acid recognition. Tetrahedron 54(14):3607–3630
10. Petersen M et al (2000) The conformations of locked nucleic acids (LNA). J Mol Recognit 13(1):44–53
11. Vester B, Wengel J (2004) LNA (locked nucleic acid): high-affinity targeting of complementary RNA and DNA. Biochemistry 43(42): 13233–13241
12. Egholm M et al (1995) Efficient pH-independent sequence-specific DNA binding by pseudoisocytosine-containing bis-PNA. Nucleic Acids Res 23(2):217–222
13. Bentin T, Larsen HJ, Nielsen PE (2003) Combined triplex/duplex invasion of double-stranded DNA by "tail-clamp" peptide nucleic acid. Biochemistry 42(47):13987–13995
14. Kaihatsu K et al (2003) Extending recognition by peptide nucleic acids (PNAs): binding to duplex DNA and inhibition of transcription by tail-clamp PNA-peptide conjugates. Biochemistry 42(47):13996–14003
15. Lohse J, Dahl O, Nielsen PE (1999) Double duplex invasion by peptide nucleic acid: a general principle for sequence-specific targeting of double-stranded DNA. Proc Natl Acad Sci U S A 96(21):11804–11808
16. Sazani P et al (2001) Nuclear antisense effects of neutral, anionic and cationic oligonucleotide analogs. Nucleic Acids Res 29(19):3965–3974
17. Koppelhus U et al (2008) Improved cellular activity of antisense peptide nucleic acids by conjugation to a cationic peptide-lipid (CatLip) domain. Bioconjug Chem 19(8):1526–1534
18. Rogers FA et al (2004) Peptide conjugates for chromosomal gene targeting by triplex-forming oligonucleotides. Nucleic Acids Res 32(22):6595–6604
19. Faria M et al (2000) Targeted inhibition of transcription elongation in cells mediated by triplex-forming oligonucleotides. Proc Natl Acad Sci U S A 97(8):3862–3867
20. Birg F et al (1990) Inhibition of simian virus 40 DNA replication in CV-1 cells by an oligodeoxynucleotide covalently linked to an intercalating agent. Nucleic Acids Res 18(10): 2901–2908
21. Maher LJ III, Wold B, Dervan PB (1989) Inhibition of DNA binding proteins by oligonucleotide-directed triple helix formation. Science 245(4919):725–730
22. Havre PA et al (1993) Targeted mutagenesis of DNA using triple helix-forming oligonucleotides linked to psoralen. Proc Natl Acad Sci U S A 90(16):7879–7883
23. Takasugi M et al (1991) Sequence-specific photo-induced cross-linking of the two strands of double-helical DNA by a psoralen covalently linked to a triple helix-forming oligonucleotide. Proc Natl Acad Sci U S A 88(13):5602–5606
24. Vasquez KM et al (1996) High-efficiency triple-helix-mediated photo-cross-linking at a targeted site within a selectable mammalian gene. Biochemistry 35(33):10712–10719
25. Wang G, Seidman MM, Glazer PM (1996) Mutagenesis in mammalian cells induced by triple helix formation and transcription-coupled repair. Science 271(5250):802–805
26. Chin JY et al (2008) Correction of a splice-site mutation in the beta-globin gene stimulated by triplex-forming peptide nucleic acids. Proc Natl Acad Sci U S A 105(36):13514–13519
27. Schleifman EB et al (2011) Targeted disruption of the CCR5 gene in human hematopoietic stem cells stimulated by peptide nucleic acids. Chem Biol 18(9):1189–1198
28. Rogers FA et al (2012) Targeted gene modification of hematopoietic progenitor cells in mice following systemic administration of a PNA-peptide conjugate. Mol Ther 20(1):109–118
29. McNeer NA et al (2012) Systemic delivery of triplex-forming PNA and donor DNA by nanoparticles mediates site-specific genome editing of human hematopoietic cells in vivo. Gene Ther 20(6):658–669
30. Yin H et al (2010) Optimization of peptide nucleic acid antisense oligonucleotides for local and systemic dystrophin splice correction in the mdx mouse. Mol Ther 18(4):819–827
31. Roberts J et al (2006) Efficient and persistent splice switching by systemically delivered LNA oligonucleotides in mice. Mol Ther 14(4): 471–475
32. Singhal G et al (2011) DNA triplex-mediated inhibition of MET leads to cell death and tumor regression in hepatoma. Cancer Gene Ther 18(7):520–530
33. Cogoi S et al (2004) Antiproliferative activity of a triplex-forming oligonucleotide recognizing a Ki-ras polypurine/polypyrimidine motif correlates with protein binding. Cancer Gene Ther 11(7):465–476
34. Shen C et al (2003) Targeting bcl-2 by triplex-forming oligonucleotide—a promising carrier for gene-radiotherapy. Cancer Biother Radiopharm 18(1):17–26
35. Taniguchi Y, Sasaki S (2012) An efficient antigene activity and antiproliferative effect by targeting the Bcl-2 or survivin gene with triplex forming oligonucleotides containing a W-shaped nucleoside analogue (WNA-betaT). Org Biomol Chem 10(41):8336–8341
36. Onyshchenko MI et al (2009) Stabilization of G-quadruplex in the BCL2 promoter region in

double-stranded DNA by invading short PNAs. Nucleic Acids Res 37(22):7570–7580
37. Ebbinghaus SW et al (1993) Triplex formation inhibits HER-2/neu transcription in vitro. J Clin Invest 92(5):2433–2439
38. Rogers FA et al (2002) Site-directed recombination via bifunctional PNA-DNA conjugates. Proc Natl Acad Sci U S A 99(26): 16695–16700
39. Lonkar P et al (2009) Targeted correction of a thalassemia-associated beta-globin mutation induced by pseudo-complementary peptide nucleic acids. Nucleic Acids Res 37(11): 3635–3644
40. McNeer NA et al (2011) Nanoparticles deliver triplex-forming PNAs for site-specific genomic recombination in CD34+ human hematopoietic progenitors. Mol Ther 19(1): 172–180
41. Vasquez KM, Narayanan L, Glazer PM (2000) Specific mutations induced by triplex-forming oligonucleotides in mice. Science 290(5491): 530–533
42. Chin JY, Schleifman EB, Glazer PM (2007) Repair and recombination induced by triple helix DNA. Front Biosci 12:4288–4297
43. Chin JY, Glazer PM (2009) Repair of DNA lesions associated with triplex-forming oligonucleotides. Mol Carcinog 48(4):389–399
44. Vasquez KM et al (2002) Human XPA and RPA DNA repair proteins participate in specific recognition of triplex-induced helical distortions. Proc Natl Acad Sci U S A 99(9):5848–5853
45. Faruqi AF et al (2000) Triple-helix formation induces recombination in mammalian cells via a nucleotide excision repair-dependent pathway. Mol Cell Biol 20(3):990–1000
46. Datta HJ et al (2001) Triplex-induced recombination in human cell-free extracts. Dependence on XPA and HsRad51. J Biol Chem 276(21): 18018–18023
47. Knauert MP et al (2005) Distance and affinity dependence of triplex-induced recombination. Biochemistry 44(10):3856–3864
48. Kim KH, Nielsen PE, Glazer PM (2006) Site-specific gene modification by PNAs conjugated to psoralen. Biochemistry 45(1):314–323
49. Christensen L et al (1995) Solid-phase synthesis of peptide nucleic acids. J Pept Sci 1(3): 175–183
50. Maurisse R et al (2010) Comparative transfection of DNA into primary and transformed mammalian cells from different lineages. BMC Biotechnol 10:9
51. Luo D et al (1999) Controlled DNA delivery systems. Pharm Res 16(8):1300–1308
52. Blum JS, Saltzman WM (2008) High loading efficiency and tunable release of plasmid DNA encapsulated in submicron particles fabricated from PLGA conjugated with poly-L-lysine. J Control Release 129(1):66–72
53. Woodrow KA et al (2009) Intravaginal gene silencing using biodegradable polymer nanoparticles densely loaded with small-interfering RNA. Nat Mater 8(6):526–533
54. Babar IA et al (2012) Nanoparticle-based therapy in an in vivo microRNA-155 (miR-155)-dependent mouse model of lymphoma. Proc Natl Acad Sci U S A 109(26):E1695–E1704
55. Fahmy TM et al (2005) Surface modification of biodegradable polyesters with fatty acid conjugates for improved drug targeting. Biomaterials 26(28):5727–5736
56. Fields RJ et al (2012) Surface modified poly(beta amino ester)-containing nanoparticles for plasmid DNA delivery. J Control Release 164(1):41–48
57. Vasquez KM et al (1999) Chromosomal mutations induced by triplex-forming oligonucleotides in mammalian cells. Nucleic Acids Res 27(4):1176–1181
58. Gunther EJ et al (1995) Mutagenesis by 8-methoxypsoralen and 5-methylangelicin photoadducts in mouse fibroblasts: mutations at cross-linkable sites induced by offoadducts as well as cross-links. Cancer Res 55(6):1283–1288
59. Schleifman EB, Chin JY, Glazer PM (2008) Triplex-mediated gene modification. Methods Mol Biol 435:175–190

Chapter 9

Synthesis of Stabilized Alpha-Helical Peptides

Federico Bernal and Samuel G. Katz

Abstract

Stabilized alpha-helical (SAH) peptides are valuable laboratory tools to explore important protein–protein interactions. Whereas most peptides lose their secondary structure when isolated from the host protein, stapled peptides incorporate an all-hydrocarbon "staple" that reinforces their natural alpha-helical structure. Thus, stapled peptides retain their functional ability to bind their native protein targets and serve multiple experimental uses. First, they are useful for structural studies such as NMR or crystal structures that map and better define binding sites. Second, they can be used to identify small molecules that specifically target that interaction site. Third, stapled peptides can be used to test the importance of specific amino acid residues or posttranslational modifications to the binding. Fourth, they can serve as structurally competent bait to identify novel binding partners to specific alpha-helical motifs. In addition to markedly improved alpha-helicity, stapled peptides also display resistance to protease cleavage and enhanced cell permeability. Most importantly, they are useful for intracellular experiments that explore the functional consequences of blocking particular protein interactions. Because of their remarkable stability, stapled peptides can be applied to whole-animal, in vivo studies. Here we describe a protocol for the synthesis of a peptide that incorporates an all-hydrocarbon "staple" employing a ring-closing olefin metathesis reaction. With proper optimization, stapled peptides can be a fundamental, accurate laboratory tool in the modern chemical biologist's armory.

Key words Stapled peptides, NMR, Protein–protein interactions, Olefin metathesis, Alpha helix

1 Introduction

Protein interactions are fundamental to the control of nearly every cellular function. The ability to manipulate these interactions is a highly desirable capability for both physiological experimentation and therapeutic benefit. Small molecules are particularly adept at targeting small hydrophobic pockets with high affinity and specificity [1]. However, the vast majority of protein interactions are mediated by relatively large and shallow interaction surfaces that are difficult to target with small molecules. Biologics, such as antibodies, are effective at recognizing large surface areas with high affinity, but their use is limited to targets that are extracellular [2].

Narendra Wajapeyee (ed.), *Cancer Genomics and Proteomics: Methods and Protocols*, Methods in Molecular Biology, vol. 1176, DOI 10.1007/978-1-4939-0992-6_9, © Springer Science+Business Media New York 2014

Thus, many intracellular signaling pathways remain largely inaccessible to meaningful real-time manipulation.

Over billions of years of evolution Mother Nature has designed precise "lock-and-key" mechanisms to ensure specificity in the tremendous variety of intermolecular biologic recognition events. The most prevalent secondary structure element within proteins is the α-helix, which is employed for numerous protein–protein and protein–DNA interactions [3–5]. First modeled by Pauling and Corey, the alpha-helix contains 3.6 residues per turn with a rise of 0.15 nm per residue allowing for maximal reinforcement of hydrogen bonding between the amide protons and carbonyl oxygens [6]. The central core minimizes free space, while the amino acid side chains are displayed outwards to form unique topographies for interaction. Small peptides that correspond to a portion of a protein involved in such an interaction can potentially serve as effective binding partners because they accurately display a true mimic of its three-dimensional interaction surface. Nevertheless, taking a single helix out of the context of the entire protein has severe consequences on its structure and biological utility. In an aqueous environment, short peptides will often lose their alpha-helical secondary structure and unfold into random coil devoid of structure [7]. The loss of structure allows peptides to achieve an extended conformation making them prime targets for protease degradation [8]. Finally, with few exceptions, peptides cannot penetrate the cell membrane [9].

Many methods have been established for the enforcement of the alpha helical conformation, but many of these approaches make use of groups that are polar or pharmacologically labile [10, 11]. Each of these obstacles for native polypeptides can be overcome by the insertion of an all-hydrocarbon cross-link [12–15]. In this strategy, peptides are synthesized containing two α-methyl, α-alkenyl glycine substitutions positioned either one or two helix turns apart followed by a ruthenium-catalyzed, ring-closing olefin metathesis reaction which is compatible with modern solid-phase peptide synthesis protocols [16, 17]. Both α,α-disubstitution and macrocyclic bridge formation promote helix formation [18]. Likewise, the resulting amphipathic helix can promote cell permeability and precludes unfolding of the peptide enhancing its proteolytic stability [19].

Methodical examination of stereochemistry and chain length revealed that amino acids containing 5 carbon atoms in their alkenyl side chain with *S* stereochemistry (S_5) placed one α-helical turn apart or, in the i and $i+4$ positions, is ideal [13]. Alternatively, to span two α-helical turns, the R stereoisomer with an 8 carbon-atom alkenyl chain (R_8) is placed at the i position and S_5 in the $i+7$ position [20].

One of the benefits of synthesizing a peptide is the flexibility to include select modifications. For example, incorporating amino acids that are phosphorylated, methylated, or glycosylated easily explores posttranslational modifications. Moreover, peptides can be easily tracked in microscopy experiments by synthesizing them with specific labels such as FITC or various Alexa Fluor dyes [12]. Addition of a biotin moiety bestows a useful handle to pull down the peptide and any interacting partners with streptavidin [21]. A photoreactive group for UV light-induced covalent cross-linking can be used to capture interacting proteins and map interaction surfaces [21, 22]. Therefore, in addition to simply blocking a specific interaction, stapled peptides are valuable tool compounds in exploration of protein–protein interaction mechanisms in general. The methods subsequently described are broadly applicable and have been used to successfully manipulate numerous and diverse protein targets.

2 Materials

1. Rink amide MBHA resin.
2. 1-Methyl-2-pyrrolidinone (NMP).
3. Piperidine.
4. 2-(6-Chloro-1*H*-benzotriazole-1-yl)-1,1,3,3-tetramethylaminium hexaflurophosphate (HCTU).
5. *N,N*-diisopropyl ethylamine (DIEA).
6. 1,2-Dichloroethane (DCE).
7. Grubbs catalyst (Grubbs I (*bis*(tricyclohexylphosphine)-benzylidene ruthenium (IV) dichloride)).
8. Dichloromethane (DCM).
9. Methanol.
10. Methyl-tert-butyl ether.
11. Acetic anhydride.
12. Trifluoroacetic acid (TFA).
13. Triisopropylsilane (TIS).
14. 1,2-Ethanedithiol (EDT).
15. Diethyl ether.
16. Acetonitrile.
17. Fmoc amino acids.
18. Fritted reaction vessel (e.g., disposable chromatography columns or peptide synthesis glassware).
19. Machines: Peptide synthesizer.
20. Machines: HPLC/LC/MS.

3 Methods

3.1 Design Considerations

In the design of stapled alpha-helical peptides, it is crucial to mimic as closely as possible the original sequence of the protein in question. Residues known or presumed to be involved in the binding interaction being targeted should remain intact. In turn, the staple should be located away from the binding interface to avoid interference with the interaction surface. Because the location and size of the staple are crucial to the successful stabilization of the helix, panels of compounds with staples at different allowed locations with different chain lengths should be designed and synthesized. Once the ideal position and size of the staple have been found (based on in vitro binding studies with its target), substitutions to the native sequence can be undertaken in order to optimize the peptide for improved binding and potency.

3.2 Peptide Synthesis

The synthesis of stapled peptides is carried out using standard Fmoc (fluorenylmethoxycarbonyl) solid-phase peptide synthesis protocols (Fig. 1). The procedure outlined below is suitable for the synthesis of a single stapled peptide in a 25 μmol scale.

1. Swell Rink Amide MBHA resin (100–200 mesh, preferably with a substitution lower than 0.4 mmol/g) with 1 mL NMP for 15 min.
2. To remove the Fmoc-protecting group, wash the resin with 1 mL of a 20 % (v/v) solution of piperidine in NMP for 15 min and drain. Repeat once.
3. Wash resin with 1 mL NMP for 1 min draining to waste. Repeat five times.
4. Couple the Fmoc amino acid onto the drained resin by addition of 10 equivalents N-α-Fmoc-protected amino acid in NMP, 9.9 equivalents of HCTU in NMP, and 20 equivalents of DIEA. The reaction is shaken at room temperature for 45 min. If coupling a sterically hindered amino acid (His, Ile, Pro, Thr, Trp, Val) shake for 60 min. To couple an olefinic amino acid, add the cross-linker, HCTU, and DIEA in a molar ratio of 4:3.8:8 to the resin and shake for 60 min.
5. Wash resin with 1 mL NMP for 1 min, five times draining to waste after each wash.
6. Go back to **step 2**, and repeat for each amino acid to add.
7. Once the synthesis has been completed, the resin is washed twice with DCM (1 mL × 3 min) and shrunk with methanol (1 mL × 5 min).

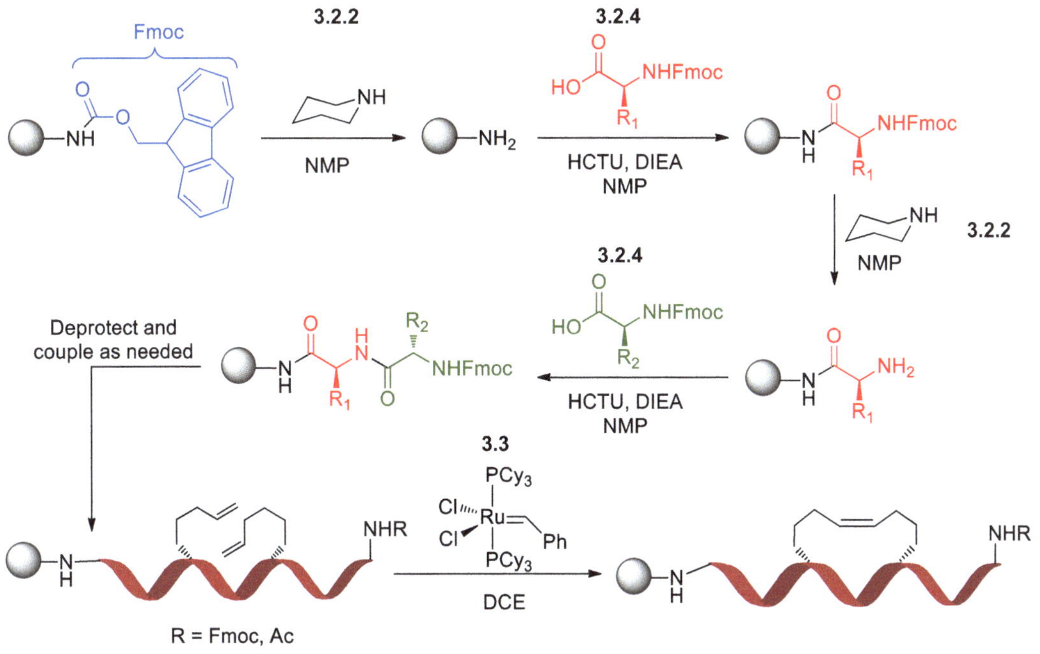

Fig. 1 Scheme for Fmoc-based solid-phase peptide synthesis with ruthenium-catalyzed olefin metathesis. Reactions are numbered by step within the protocol, where Subheading 3.2, **step 2** represents the deprotection reaction, Subheading 3.2, **step 4** the coupling reaction, and Subheading 3.3 the metathesis reaction. The chemical structure in Subheading 3.2, **step 2** is that of piperidine, Subheading 3.2, **step 4** the Fmoc-protected amino acid, and Subheading 3.3 the Grubbs catalyst. *Gray ball* represents rink amide MBHA resin

3.3 Olefin Metathesis

In order to carry out the olefin metathesis reaction, the *N*-terminus of the peptide must not be exposed as the free amine. The reaction may be carried out with the peptide protected as the Fmoc carbamate.

1. Swell resin by adding 1 mL dry DCE for 30 min.
2. Add 1 mL of freshly dissolved 10 mM Grubbs catalyst in DCE for 2 h with constant bubbling under nitrogen. Add additional DCE to avoid loss of volume due to evaporation.
3. Wash once with DCE (1 mL × 1 min). Repeat **step 2**.
4. Wash three times with DCE (1 mL × 1 min).
5. Shrink resin with methanol (1 mL × 5 min).

3.4 N-Terminus Modifications

1. The *N*-terminus of the peptide can be modified using amine-reactive agents. Acetylation of the *N*-terminus is carried out by treating the pre-swollen and deprotected (i.e., free amine at the *N*-terminus) resin with 1 mL of a 2:1 (v/v) mixture of acetic anhydride and DIEA.
2. To attach other modifications such as fluoresceination or biotinylation, the swollen and deprotected resin is exposed to fluorescein isothiocyanate (FITC) or biotinyl-*O*-succinimide (Biotin-OSu) in the presence of DIEA.

3.5 Cleavage and Deprotection

1. Freshly make a cleavage cocktail that is 95 % TFA, 2.5 % TIS, and 2.5 % water on ice (CAUTION: mixture is exothermic). If the peptide has sulfur-containing amino acids or modifications (e.g., cysteine, methionine, biotin) then use 94 % TFA, 1 % TIS, 2.5 % water, and 2.5 % EDT.
2. Add 1 mL of cleavage cocktail per 20 mg peptide-bound resin, and shake at room temperature for 2–3 h (NOTE: the longer reaction times are necessary for peptides containing multiple arginines).
3. Filter the reaction mixture through a fritted disposable chromatography column, and collect the filtrate in a chilled, 15 mL conical tube containing 5–6 mL of cold diethyl ether. The peptide should precipitate immediately upon contact with the solvent.
4. Pellet the peptide by centrifugation at 3,000 × *g* for 15 min at 4 °C.
5. Decant the ether supernatant, and air-dry the pelleted stapled peptide.

3.6 Purification

1. Dissolve the peptide in approximately 1 mL of 50 % acetonitrile in water.
2. Inject the peptide on a reverse-phase high-performance liquid chromatography with a C18 column and a mobile-phase gradient of water and acetonitrile, each with 0.1 % TFA.
3. Monitor the HPLC fractions by LC/MS, pooling fractions.
4. Pooled fractions are lyophilized overnight.
5. The lyophilized peptide is dissolved in DMSO and quantified by amino acid analysis.
6. Prepare stock solutions in DMSO at 1–10 mM and store at 4 °C or −20 °C.

4 Notes

1. Be sure to take proper precautions before using any chemicals. This includes appropriate safety goggles, lab coats, and closed toe shoes. All reactions should be carried out in a chemical fume hood.
2. Piperidine should be aliquoted using a disposable syringe and needle as the acidic fumes will corrode pipettes.
3. Make sure that the frit is firmly in place and flush against the bottom before adding any reagents.
4. Optimization of the peptide through iterative syntheses that explore alternative staple sties and start and stop positions is

essential in the process. These factors can be modified in order to avoid helix breakers (e.g., proline and glycine). Each compound should be tested for helicity, mechanism of action, and cell permeability.

5. In general, sulfur-containing amino acids should be avoided because they may inactivate the ruthenium-containing Grubbs catalyst, leading to potentially low yields of product. If the use of a cysteine is necessary, then the trityl-protected variant Fmoc-Cys(Tr)-OH should be employed during coupling. Methionines can be substituted with the isosteric amino acid norleucine.
6. It is important to design a negative control stapled peptide to confirm target-specific effects. Ensure that the negative control mutation does not appreciably affect helicity or cell penetration.

References

1. Hopkins AL, Groom CR (2002) The druggable genome. Nat Rev Drug Discov 1(9):727–730, PubMed PMID: 12209152
2. Buss NA, Henderson SJ, McFarlane M, Shenton JM, de Haan L (2012) Monoclonal antibody therapeutics: history and future. Curr Opin Pharmacol 12(5):615–622, PubMed PMID: 22920732
3. Guharoy M, Chakrabarti P (2007) Secondary structure based analysis and classification of biological interfaces: identification of binding motifs in protein-protein interactions. Bioinformatics 23(15):1909–1918, PubMed PMID: 17510165
4. Jochim AL, Arora PS (2009) Assessment of helical interfaces in protein-protein interactions. Mol BioSyst 5(9):924–926, PubMed PMID: 19668855
5. Jones S, Thornton JM (1996) Principles of protein-protein interactions. Proc Natl Acad Sci U S A 93(1):13–20, PubMed PMID: 8552589. Pubmed Central PMCID: 40170
6. Pauling L, Corey RB, Branson HR (1951) The structure of proteins; two hydrogen-bonded helical configurations of the polypeptide chain. Proc Natl Acad Sci U S A 37(4):205–211, PubMed PMID: 14816373. Pubmed Central PMCID: 1063337
7. Kéri GR, Tóth IN. (2003) Molecular pathomechanisms and new trends in drug research. London; New York: Taylor & Francis xiv, 635
8. Tyndall JD, Nall T, Fairlie DP (2005) Proteases universally recognize beta strands in their active sites. Chem Rev 105(3):973–999, PubMed PMID: 15755082
9. Madani F, Lindberg S, Langel U, Futaki S, Graslund A (2011) Mechanisms of cellular uptake of cell-penetrating peptides. J Biophys 2011:414729, PubMed PMID: 21687343. Pubmed Central PMCID: 3103903
10. Garner J, Harding MM (2007) Design and synthesis of alpha-helical peptides and mimetics. Organic Biomol Chem 5(22):3577–3585, PubMed PMID: 17971985
11. Henchey LK, Jochim AL, Arora PS (2008) Contemporary strategies for the stabilization of peptides in the alpha-helical conformation. Curr Opin Chem Biol 12(6):692–697, PubMed PMID: 18793750. Pubmed Central PMCID: 2650020
12. Walensky LD, Pitter K, Morash J, Oh KJ, Barbuto S, Fisher J et al (2006) A stapled BID BH3 helix directly binds and activates BAX. Mol Cell 24(2):199–210, PubMed PMID: 17052454
13. Schafmeister CE, Po J, Verdine GL (2000) An all-hydrocarbon cross-linking system for enhancing the helicity and metabolic stability of peptides. J Am Chem Soc 122(24):5891–5892, PubMed PMID: WOS:000087845700030. English
14. Brown CJ, Cheok CF, Verma CS, Lane DP (2011) Reactivation of p53: from peptides to small molecules. Trends Pharmacol Sci 32(1): 53–62, PubMed PMID: 21145600
15. Meyers RA (2004) Encyclopedia of molecular cell biology and molecular medicine, 2nd edn. Weinheim, Wiley-VCH Verlag
16. Blackwell HE, Grubbs RH (1998) Highly efficient synthesis of covalently cross-linked peptide helices by ring-closing metathesis.

Angew Chem Int Edit 37(23):3281–3284, PubMed PMID: WOS:000077806300017. English

17. Blackwell HE, Sadowsky JD, Howard RJ, Sampson JN, Chao JA, Steinmetz WE et al (2001) Ring-closing metathesis of olefinic peptides: design, synthesis, and structural characterization of macrocyclic helical peptides. J Organic Chem 66(16):5291–5302, PubMed PMID: 11485448
18. Venkatraman J, Shankaramma SC, Balaram P (2001) Design of folded peptides. Chem Rev 101(10):3131–3152, PubMed PMID: 11710065
19. Shepherd NE, Hoang HN, Abbenante G, Fairlie DP (2005) Single turn peptide alpha helices with exceptional stability in water. J Am Chem Soc 127(9):2974–2983, PubMed PMID: 15740134
20. Bernal F, Tyler AF, Korsmeyer SJ, Walensky LD, Verdine GL (2007) Reactivation of the p53 tumor suppressor pathway by a stapled p53 peptide. J Am Chem Soc 129(9):2456–2457, PubMed PMID: 17284038
21. Braun CR, Mintseris J, Gavathiotis E, Bird GH, Gygi SP, Walensky LD (2010) Photoreactive stapled BH3 peptides to dissect the BCL-2 family interactome. Chem Biol 17(12):1325–1333, PubMed PMID: 21168768. Pubmed Central PMCID: 3048092
22. Leshchiner ES, Braun CR, Bird GH, Walensky LD (2013) Direct activation of full-length proapoptotic BAK. Proc Natl Acad Sci U S A 110(11):E986–E995, PubMed PMID: 23404709. Pubmed Central PMCID: 3600461

Chapter 10

Arginine-Grafted Biodegradable Polymer: A Versatile Transfection Reagent for both DNA and siRNA

Jagadish Beloor, Hye Yeong Nam, Sang-Kyung Lee, and Priti Kumar

Abstract

Effective delivery of DNA or siRNA into primary cells demands an efficient delivery system. However, the significant differences in physical and molecular characteristics of the two molecules generally necessitate distinct delivery systems or considerable differences in carrier formulation protocols for effective transfection. Arginine-grafted bioreducible poly (disulfide amine) (ABP) is a redox-sensitive, bioreducible, positively charged polymer which complexes with siRNA and DNA via charge interactions to form nanoplexes. ABP effectively mediates cytoplasmic delivery of both DNA and siRNA into multiple cell types, including primary cells like myoblast, human umbilical vein endothelial cells (HUVECs), and primary rat aorta vascular smooth muscle cells (SMCs) eliciting functional activity. In this chapter, we provide the detailed protocols for the synthesis of ABP as well as transfection of both siRNA and DNA using ABP.

Key words Gene delivery, siRNA delivery, Bioreducible polymers, ABP, Transfection, Redox-sensitive polymers

1 Introduction

The paucity of effective delivery agents for siRNA and plasmid DNA has hugely limited the clinical development of nucleic acid bio-drugs as therapeutics. Although viral delivery vectors are efficient, concerns of immunogenicity, random genome integration, carcinogenicity, and unnecessary long-term maintenance, specially in the case of integrating vectors, highlight the acute need for nonviral delivery systems [1, 2]. Polymer-based delivery systems have the potential to overcome these problems. In addition, the incorporation functional groups on the polymer backbone can enable dynamic changes in their physiochemical properties inducing bio-responsiveness. Thus, cleavage of covalent bonds, disassembly of noncovalent interactions, changes of protonation, conformation, or hydrophilicity/lipophilicity in response to the microenvironments in cells/tissues can lead to shielding,

Narendra Wajapeyee (ed.), *Cancer Genomics and Proteomics: Methods and Protocols*, Methods in Molecular Biology, vol. 1176, DOI 10.1007/978-1-4939-0992-6_10, © Springer Science+Business Media New York 2014

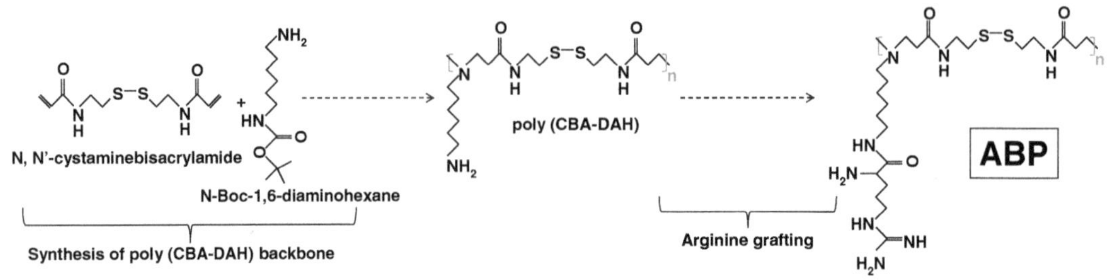

Fig. 1 Schematic depiction of ABP synthesis

deshielding, cell targeting, cellular entry, endosomal/cytosolic release, and functional accessibility of nucleic acids encapsulated within polymers [3].

For instance, the introduction of bioreducible (redox-sensitive) disulfide components in some polymers positively contributed towards the minimization of toxicity and maximization of siRNA or gene functionality [4–6].

Arginine-grafted bioreducible poly (disulfide amine) (ABP) is a unique polymer which can form nanoparticles with both siRNA and DNA and deliver them in a functionally active state into the cytosol. The synthesis of ABP and its backbone is convenient via commercially available monomers, is simple with high consistency, allows large-scale production, and is easy to handle [7–9]. ABP is created by engrafting arginine, a positively charged amino acid, onto a redox-sensitive polymer made of monomeric (CBA-DAH) subunits [10] (Fig. 1). The cationic charge of ABP enables nanoplex formation with siRNA and DNA, and the net positive charge increases with the increasing ratios of polymer to the nucleic acid. The high transfection efficiency of ABP is attributable to multiple features. The arginine moieties play an important role for cellular membrane penetration after endocytosis as in cell-penetrating peptides (CPPs) or protein transduction domains (PTDs) such as HIV-1 Tat sequence, penetratin, or oligoarginine [11]. The presence of bioreducible internal disulfide bonds in the ABP backbone leads to degradation of ABP polyplexes in the endosome, and cytoplasm not only triggers release of genetic materials but also highly reduces cytotoxicity of the polymer. The parental backbone of poly (CBA-DAH) has endosome-buffering moieties that aid in the efficient endosomal escape of the nanoplexes [7].

We and our collaborators have successfully used ABP in several formats by itself as well as a component of other viral and nonviral delivery systems for enhancing transfection efficiencies, in vitro as well as in vivo, for siRNA as well as gene delivery [7, 12–14]. In this chapter we detail the stepwise protocol for the synthesis, characterization, and optimization of ABP for delivery of siRNA or plasmid DNA.

2 Materials

2.1 Requirements for Synthesis of ABP

Prepare all solutions using ultrapure water (with a sensitivity of 18 MΩ at 25 °C obtained using Milli-Q Integral Water Purification System) and analytical grade reagents. If there is no specific indication, store all reagents at room temperature.

1. *tert*-Butyl-*N*-(6-aminohexyl) carbamate (*N*-Boc-1,6-diaminohexane, *N*-Boc-DA) (Sigma): Store at −15 °C.
2. Trifluoroacetic acid (TFA) (Sigma).
3. Triisobutylsilane (TIS) (Sigma).
4. Triisopropylsilane (TIPS) (Sigma).
5. *N,N′*-Cystaminebisacrylamide (CBA) (PolySciences, Inc., Warrington, PA, USA): Store at −15 °C.
6. 2-(1H-benzotriazole-1-yl)-1, 1, 3, 3-tetramethyluronium hexafluorophosphate (HBTU) (Novabiochem, San Diego, CA, USA): Store at 4 °C.
7. N-alpha-(9-fluorenylmethyloxycarbonyl)-*N′*-2,2,4,6,7-pentamethyldihydrobenzofuran-5-sulfonyl-L-arginine (Fmoc-L-Arg(pbf)-OH) (Anaspec, Inc., San Jose, CA, USA): Store at 4 °C.
8. Anhydrous DMF (Sigma).
9. BD Falcon™ 50 mL conical centrifuge tubes (BD, Franklin Lakes, NJ, USA).
10. Sorvall* Primo/Primo* R Benchtop Centrifuges (Thermo Scientific, Asheville, NC, USA).
11. Dialysis membrane (MWCO = 1000, Spectrum Laboratories, Inc., Rancho Dominguez, CA, USA): Store at 4 °C.
12. Syringe filter: Millex-HV Filter, 0.45 μm, PVDF, 33 mm, gamma sterilized (Millipore, Billerica, MA, USA).
13. Freeze dryer (Labconco, Kansas City, MO, USA).

2.2 Requirements for NMR Spectrum of ABP

1. Deuterium oxide (D_2O) (Sigma).
2. NMR tube: 400 MHz standard series NMR tube (NORELL, Inc., Landisville, NJ, USA).
3. 400 NMR spectrometers (Varian, Inc., Palo Alto, CA, USA).

2.3 Requirements for Molecular Weight Determination of ABP

1. Eluent buffer: 0.1 M Acetate buffer (30 % acetonitrile, v/v, pH 6.5).
2. Size-exclusion chromatography (SEC): AKTA FPLC system with a Superose 12 column and refractive index (RI) and ultraviolet (UV) detectors (GE Healthcare, Piscataway, NJ, USA).
3. Standard: Poly [*N*-(2-hydroxypropyl)-methacrylamide] (pHPMA) (GE Healthcare, Piscataway, NJ, USA).

2.4 Requirements for Electrophoretic Mobility Shift Assay

1. Cleaned electrophoretic tank containing 1× TAE (Tris–acetate–EDTA) buffer (40 mM Tris, 20 mM acetic acid, and 1 mM EDTA).
2. 1.5 % Agarose gel for siRNA or 0.7 % agarose gel for pDNA prepared in 1× TAE buffer.
3. 1.5 mL Microcentrifuge tubes.
4. Sterile phosphate-buffered saline (PBS) pH 7.4.
5. siRNA reconstituted in nuclease-free water (1 μg/μL) (*see* **Note 1**).
6. Plasmid DNA (pDNA) construct dissolved in autoclaved distilled water (500 ng/μL) (*see* **Note 1**).
7. ABP polymer dissolved in ultrapure autoclaved distilled water (*see* **Notes 2** and **3**).
8. Gel loading dye, 5× (QIAGEN).
9. UV transilluminator or gel documentation system (Kodak Gel Logic 100 imaging system).

2.5 Requirements for Cell Culture

1. Cell line of interest.
2. Complete media containing Dulbecco's modified Eagle medium (DMEM) containing 10 % fetal bovine serum and 1 % penicillin–streptomycin antibiotic solution (Invitrogen).
3. PBS, pH 7.4.
4. T-75 cm2 cell culture flasks (BD Falcon).
5. Trypsin–EDTA (Invitrogen).
6. 15 mL Polypropylene tubes.
7. Hemocytometer.
8. Inverted microscope.
9. Carbon dioxide incubator.

2.6 Requirements for Cytotoxicity Assay

1. Cell line of interest.
2. 96-Well tissue culture plates.
3. Sterile PBS (pH 7.4).
4. Cell counting kit—8 (Dojindo Molecular Technologies).
5. Scrambled siRNA targeting luciferase gene (siLuci) (*see* **Note 4**).
6. ABP polymer (*see* **Notes 2** and **3**).
7. pDNA construct dissolved in autoclaved distilled water (500 ng/μL) (*see* **Note 1**).
8. Microcentrifuge tubes.
9. Carbon dioxide incubator.
10. Microplate reader with 450 nm filters.

2.7 Requirements for Transfection

1. 12-Well cell culture plates.
2. Cells of interest.
3. Complete media containing DMEM containing 10 % fetal bovine serum and 1 % penicillin–streptomycin antibiotic solution.
4. ABP.
5. pDNA of interest (a reporter construct expressing a marker gene like GFP is a good test for efficiencies).
6. siRNAs for the target gene—siRNA targeting murine superoxide dismutase-1 (siSOD1) (*see* **Note 5**).
7. Microcentrifuge tubes.
8. Autoclaved sterile PBS (pH 7.4).

3 Methods

3.1 Synthesis of Poly (CBA-DAH) Backbone

1. Add 0.216 g *N*-Boc-DAH, 0.260 g of CBA, and 1 mL of 10 % aqueous methanol solution into a 100 mL two-neck flask, stir the mixture, and purge the solution with nitrogen gas (*see* **Note 5**).
2. Heat and stir the solution at 60 °C for 3 days.
3. Add 0.021 g of *N*-Boc-DAH into the reaction mixture, and carry out the reaction at 60 °C for an additional 2 h (*see* **Note 6**).
4. Add the mixture solution into two 40 mL of cold diethyl ether in 50 mL falcon tubes. Shake solution vigorously and collect using centrifugation at 4 °C.
5. Add 1 mL of DMF in pellets, and vortex until clear. Repeat **steps 4** and **5** three times, and obtain product pellets (*see* **Note 7**).
6. Remove ether solvent from pellets using N_2 bubbling, and combine all pellets.
7. Add 3 mL of TFA/TIS/H_2O mixture (2.85 mL:0.075 mL: 0.075 mL) in 50 mL Falcon tube with pellets, and stir vigorously at ice bath for 30 min (*see* **Note 8**).
8. Repeat **steps 4**, **5**, and **6**.
9. Add water until pellets are dissolved completely with shaking, and carry out dialysis with dialysis membrane against ultrapure water overnight. Change water four to five times during dialysis.
10. Filter aqueous solution with product using syringe filter, freeze, and lyophilize using freeze dryer.

3.2 Procedure for Arginine Grafting

1. Add 0.476 g of poly (CBA-DAH) in a 100 mL round-bottomed flask, and dissolve them with 10 mL of DMF.
2. Add 2.80 g of Fmoc-Arg(pbf)-OH, 1.52 g of HBTU, and 1.4 mL of DIPEA into solution 1, and stir the solution. Carry out the reaction at room temperature for 2 days (*see* **Note 9**).

3. Add the mixture solution into two 40 mL of cold diethyl ether in 50 mL Falcon tubes. Shake solution vigorously and collect using centrifugation at 4 °C.
4. Add 1 mL of DMF in pellets, and vortex until clear. Repeat **steps 4** and **5** three times, and obtain product pellets (*see* **Note 7**).
5. Remove ether solvent from pellets using N_2 bubbling, and combine all pellets.
6. Dissolve pellets in 50 mL Falcon tube with 3 mL of DMF, and mix the solution with 3 mL of piperidine (0.9 mL) solution in DMF (2.1 mL) at room temperature for 30 min (*see* **Note 10**).
7. Repeat **steps 3**, **4**, and **5**.
8. Add 3 mL of TFA/TIPS/H_2O mixture (2.85 mL:0.075 mL: 0.075 mL) to obtain products in Falcon tube from **step 7**, and stir vigorously at ice bath for 30 min (*see* **Note 11**).
9. Repeat **steps 3**, **4**, and **5**.
10. Add water until pellets are dissolved completely with shaking, and carry out dialysis with dialysis membrane against ultrapure water overnight. Change water four to five times during dialysis.
11. Filter aqueous solution with product using syringe filter, freeze, and lyophilize using freeze dryer (*see* **Note 12**).

3.3 Confirmation Using NMR

1. Weight about 5 mg of each poly (CBA-DAH) and ABP, and dissolve them in 0.6 mL of D_2O.
2. Transfer each solution to NMR tube, and obtain spectrum data using the Mercury 400 NMR spectrometers.
3. Running and analysis of the NMR and SEC are performed by a professional operator.

The synthesis of poly (CBA-DAH) is confirmed by proton NMR spectrum through the disappearance of signal peaks between δ 5 and 7 ppm, indicating that the acrylamide end groups do not exist in the final poly (CBA-DAH) [10]. The arginine modification is confirmed by evaluating the ratio between the proton peaks of integrated arginine and integrated poly (CBA-DAH) [7].

3.4 Molecular Weight Determination and Quantification of ABP Using HPLC Method

1. Weight about 5 mg of each polymer, and dissolve them in 1 mL of eluent buffer.
2. Inject samples into the SEC, and elute the elution buffer.
3. Analyze the spectrum using calibration curve with pHPMA standards.
4. The molecular weight of poly (CBA-DAH) and ABP is expected to be about 4.5 kDa (PDI = 1.5) and 3.6 kDa (PDI = 1.2).

3.5 Protocol for EMSA

1. Label the microcentrifuge as (nucleic acid:ABP weight ratios) (a) 1:0, (b) 1:1, (c) 1:3, (d) 1:5, (e) 1:10, and (f) 1:20.
2. Prepare ABP concentrations of 2 mg/mL (for siRNA) or 1 mg/mL (for pDNA) experiments, respectively, from the stocks (*see* **Notes 13–16**).
3. Pipette out 0 μL, 0.5 μL, 1.5 μL, 2.5 μL, 5 μL, and 10 μL of ABP polymer (as diluted above), respectively, to the above labelled tubes (*see* **Notes 2** and **3**).
4. Add 1 μL of nucleic acid (siRNA or pDNA) to each tube containing different amounts of ABP.
5. Make up the final volume to 20 μL by adding 19 μL, 18.5 μL, 17.5 μL, 16.5 μL, 14.5 μL, and 9 μL of PBS pH 7.4, respectively.
6. Mix the contents of the tube by tapping or pipetting two to three times, and incubate the tube at room temperature for 30 min to form nucleic acid/ABP polyplexes.
7. Load the polyplexes onto 1.5 % (for siRNA) or 0.7 % (for pDNA) agarose gel mixed with 5 μL gel loading dye, and perform the electrophoresis at 135 V for 15 min (for siRNA) or 100 V for 30 min (for pDNA) (Fig. 2).

3.6 Protocol for Cell Culture

1. Prepare complete media by adding 50 mL of FBS and 5 mL of penicillin–streptomycin antibiotics to 445 mL to DMEM. Mix the contents properly.
2. Warm the complete media in 37 °C water bath for 15–30 min prior to cell culture.
3. Simultaneously thaw the trypsin–EDTA aliquots.
4. Check for the confluency of cells in T-75 culture flask and the condition of cells under microscope (*see* **Note 17**). Continue with the next step if the confluency of the cells has reached 80–90 % and the cells are in good condition.
5. Wash the cells with 10 mL of PBS (pH 7.4), and aspirate the PBS using an aspirator.

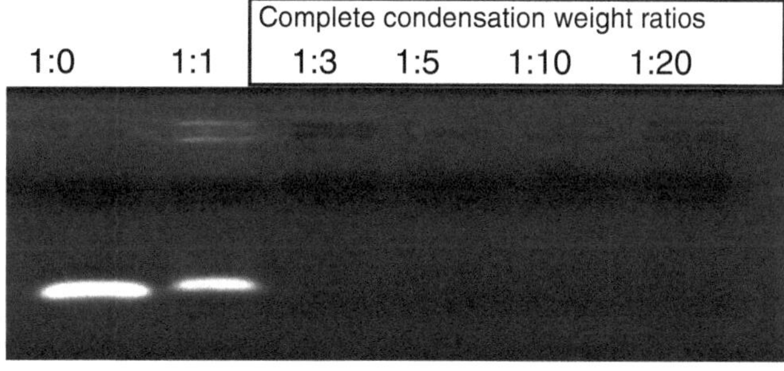

Fig. 2 Example of an EMSA for condensation of ABP with siRNA at different weight ratios

6. Add 2 mL of trypsin–EDTA, and incubate the flask in CO_2 incubator for 3–5 min. Check for complete detachment of cells from the flask.
7. Add 8 mL of complete media, and harvest cells into 15 mL polypropylene tubes.
8. Recover the cells by centrifuging at 1,200 rpm for 3 min. Aspirate the supernatant, and wash the cells with PBS (pH 7.4).
9. Count the cells using hemocytometer.

3.7 Protocol for Cytotoxicity Assay

Determination of nontoxic weight ratios specific for the cell line of interest is very important. Selection of weight ratios for further identifying the best transfection ratio should range from the weight ratio that induces complete condensation of siRNA or pDNA (Fig. 1) to maximum weight ratio that does not have any cytotoxicity (Fig. 2) with respect to cell line of interest.

1. Seed 1×10^4 cells/well in a 96-well plate 24 h prior to treatment with siRNA/ABP polyplex in triplicates.
2. Next day, approximately after 24 h, wash the cells with sterile PBS (pH 7.4) and add 200 μL of plain DMEM (without FBS or antibiotics) to each well. 1 h prior to transfecting the cells place the plate in incubator.
   (a) Prepare the polyplexes with different weight ratios of ABP (1:0, 1:5, 1:10, 1:20, 1:40, 1:60), and label the tubes accordingly in triplicates.
   (b) Firstly prepare 500 ng/μL siRNA or pDNA working concentrations from their respective stocks, and also dilute the ABP to 2.5 μg/μL concentration (*see* **Notes 2**, **3**, and **18**).
   (c) Pipette out 1 μL of nucleic acid to each tube, and add 0, 1, 2, 4, 8, and 12 μL of ABP (2.5 μg/μL) to respective tubes.
   (d) Make up the volume to 10 μL by adding 9, 8, 7, 5, 1, and 0 μL of sterile autoclaved PBS (pH 7.4) to the respective triplicate tubes.
   (e) Mix the contents of tube by pipetting two to three times and incubate at room temperature for 30 min.
3. Remove the plain media from the culture plate, treat 10 μL polyplexes to the cells in triplicate wells, and immediately add 90 μL of plain DMEM. Also maintain untreated mock cells in triplicates.
4. After 4 h of incubation in CO_2 incubator, wash cells with sterile autoclaved PBS (pH 7.4). Add 200 μL of complete DMEM media, and incubate for the next 20 h.
5. Next day, 10 μL of CCK-8 solution is added to each well containing the media and incubate for 1–4 h in the CO_2 incubator (*see* **Notes 19** and **20**).

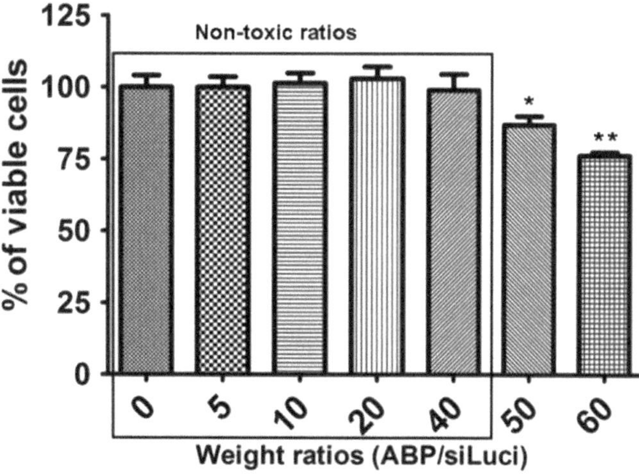

Fig. 3 Determination of cytotoxic weight ratios of ABP complexed with a siRNA targeting the firefly luciferase in B16-F10 cells. Here, weight ratios of 1:50 and beyond were significantly (* $p<0.05$, ** $p<0.01$) toxic to B16-F10 cells and hence weight ratios of 1–40 would be chosen for the experiment

6. Take the absorbance at 450 nm using a microplate reader (*see* **Note 21**).
7. Consider mock treatment readings as 100 % viability, and accordingly calculate the percentage of cell viability in other treatments with respect to mock (Fig. 3).

3.8 Protocol for Transfection

1. Seed 1×10^5 cells/well in a 12-well plate 24 h prior to transfection with siRNA/ABP polyplex.
2. Twenty-four hours post-seeding wash the cells with sterile PBS (pH 7.4), add 500 μL of plain DMEM (without FBS or antibiotics) to each well, and incubate the cells for 1 h prior to transfection.
3. Meanwhile, prepare the polyplexes with increasing weight ratios of ABP (1:0, 1:5, 1:10, 1:20, and 1:40).
   (a) Label the microcentrifuge tubes as mock, 1:0, 1:5, 1:10, 1:20, and 1:40, for siRNA transfection, and for pDNA transfection label tubes as mock, 1:0, 1:5, 1:10, 1:20, 1:40, and 1:50.
   (b) Thaw the siRNA or the pDNA (500 ng/μL) and ABP (5 μg/μL) stocks on ice.
   (c) Pipette out 90 μL of sterile PBS (pH 7.4) into each tube, and add 2 μL of siRNA or pDNA to the tubes.
   (d) Add 0 μL, 1 μL, 2 μL, 4 μL, 8 μL, and 10 μL (for siRNA) or 0 μL, 1 μL, 2 μL, 4 μL, 8 μL, 10 μL, and 10 μL (for pDNA) of ABP, respectively, to the above labelled tubes

(**step a**), and mix by pipetting the contents three to five times.

(e) Incubate the tube for 30 min at room temperature.

4. Remove the plain media from the wells and transfect with polyplexes prepared by dropping all over the well. Immediately add 400 μL of plain DMEM (without FBS or antibiotics).
5. After 4 h of incubation in CO_2 incubator (*see* **Notes 22** and **23**), wash cells with sterile autoclaved PBS (pH 7.4). Add 1 mL of complete DMEM media, and incubate for the next 20 h (to analyze gene silencing by siRNA) or 44 h (for gene expression by pDNA).
6. Next day, after 24 h of transfection with siRNA, isolate the total RNA, prepare cDNA, and check for gene silencing by standard RT-PCR or real-time PCR using specific primers. If using fluorescently labelled siRNA (for example siRNA labelled with FITC), analyze siRNA uptake using standard flow cytometry.
7. After 48 h of transfection with pDNA, harvest the cells by trypsinization and analyze the GFP expression using standard flow cytometry.

4 Notes

1. While performing electrophoretic mobility shift assay (EMSA) or working with siRNA or DNA, it is important to maintain nuclease-free condition.
2. ABP polymer is composed of disulfide bonds which are sensitive to reducing agents such as dithiothreitol (DTT) or beta-mercaptoethanol (BME), and exposing ABP to these agents leads to degradation into smaller pieces rendering the polymer incapable of complexing with nucleic acids.
3. As ABP is a water-soluble polymer, dissolve the polymer at 5 μg/μL concentrations as stock, store at –20 °C, and dilute to required concentrations when necessary.
4. siLuci sequence 5′-GGACAUUACUAGUGACUCA-3′.
5. The reaction should be performed in the dark and under nitrogen gas.
6. 10 % Molar excess of *N*-Boc-DAH is added into the reaction solution to stop the polymerization.
7. Precipitation with ether is performed to remove unreacted starting materials and impurities.
8. *N*-Boc-protecting group is removed after reaction with a TFA/TIS/H_2O mixture (95/2.5/2.5 v/v). Pay attention to dissolve pellets completely.

9. For this coupling reaction, Fmoc-Arg(pbf)-OH (4 equiv.), HBTU (4 equiv.), and DIPEA(8 equiv.) are added to poly (CBA-DAH) in DMF.
10. An equal volume of piperidine solution (30 % in DMF) is added to the crude product solution in DMF to remove Fmoc protection groups of arginine residues.
11. The reagent solution (TFA:TIPS:H_2O = 95: 2.5: 2.5, v/v) is used for the deprotection of pbf groups in arginine residues.
12. The synthesized poly (CBA-DAH) and ABP powder after lyophilization are stored in a sealed tube filled with nitrogen gas at −80 °C.
13. siRNAs are usually 19–21 nucleotide size; thus most of the siRNAs designed have a very similar charge. While performing EMSA one can use any siRNA sequence to identify the weight ratio of the polymer which results in complete condensation of siRNA.
14. siRNAs are usually provided as lyophilized pellets, so resuspend the pellet in nuclease-free water at a stock concentration of 100 pmol/μL (100 μM), store in 10–50 μL in microcentrifuge tubes, and store in −70 °C. In general, 100 pmol siRNA weighs approximately about 1.35 μg; hence weight ratios of ABP can be calculated accordingly.
15. Avoid repeated freeze thawing of siRNA and DNA stocks as it may lead to degradation and affect the transfection efficiency.
16. In case of EMSA for pDNA, it is always better to check the complete condensing weight ratio with the specific pDNA to be used for transfection as the overall charge imparted might vary based on the size of pDNA.
17. Freshly thawed cells from the liquid nitrogen-stored stocks should undergo at least three passages before being used for cytotoxicity or transfection studies. This ensures the stability of cells and removal of dead or weak cells from the plate which contribute towards the false cytotoxicity or nonspecific uptake of the ABP polyplex.
18. While assessing cytotoxicity with siRNA/ABP complexes, a known nontoxic siRNA should be used; this helps to avoid any nonspecific cytotoxicity contributed by the silencing of host gene. This can be a scrambled sequence derived from the siRNA of interest or a siRNA targeting a gene like the firefly luciferase gene or green fluorescent protein that is not present in mammalian cells.
19. While adding CCK-8 reagent to the culture wells avoid bubble formation, as these bubbles interfere with the O.D. readings.
20. Incubation time for color development depends on the type of cells and the cell density in the culture well. Hence the time

incubation time period has to be determined in particular to the cell type being tested.

21. The readings can be taken immediately, or to measure later the reaction can be stopped by adding 10 μL 0.1 M HCl to each well, and the reading can be taken at any point until the next 24 h without any change. The plate has to be stored in dark condition at room temperature.
22. Look for the condition of the cells; if cells appear healthy and stable extending the incubation time up to 6 h will enhance the transfection efficiency.
23. Transfection of pDNA can also be performed in the presence of serum. Although the presence of serum does not affect the transfection efficiency significantly, serum-free plain media may enhance efficiencies for siRNA transfections.

Acknowledgements

This work was supported by the Korea National Research Foundation Grant (2012K0001414).

References

1. Luo D, Saltzman WM (2000) Synthetic DNA delivery systems. Nat Biotechnol 18:33–37
2. Nayerossadat N, Maedeh T, Ali PA (2012) Viral and nonviral delivery systems for gene delivery. Adv Biomed Res 1:27
3. Edinger D, Wagner E (2011) Bioresponsive polymers for the delivery of therapeutic nucleic acids. Nanomed Nanobiotechnol 3: 33–46
4. Kim T, Kim SW (2011) Bioreducible polymers for gene delivery. React Funct Polym 71: 344–349
5. Liu F, Huang L (2002) Development of non-viral vectors for systemic gene delivery. J Control Release 78:259–266
6. Son S, Namgung R, Kim J, Singha K, Kim WJ (2012) Bioreducible Polymers for Gene Silencing and Delivery. Acc Chem Res 45:1100–1112
7. Kim T, Ou M, Lee M, Kim SW (2009) Arginine-grafted Bioreducible Poly(disulfide amine) for Gene Delivery Systems. Biomaterials 30:658–664
8. Danusso F, Ferruti P (1970) Synthesis of tertiary amine polymers. Polymer 11:88–113
9. Ferruti P, Marchisio MA, Barbucci R (1985) Synthesis, physico-chemical properties and biomedical applications of poly(amido-amine) s. Polymer 26:1336–1348
10. Ou M, Wang X, Xu R, Chang C, Bull DA, Kim SW (2008) Novel Biodegradable Poly(disulfide amine)s for Gene Delivery with High Efficiency and Low Cytotoxicity. Bioconjug Chem 19: 626–633
11. Futaki S (2005) Membrane-permeable arginine-rich peptides and the translocation mechanisms. Adv Drug Deliv Rev 57:547–558
12. Kim J, Kim PH, Nam HY, Lee JS, Yun CO, Kim SW (2012) Linearized oncolytic adenoviral plasmid DNA delivered by bioreducible polymers. J Control Release 158:451–460
13. Beloor J, Choi CS, Nam HY et al (2012) Arginine-engrafted biodegradable polymer for the systemic delivery of therapeutic siRNA. Biomaterials 33:1640–1650
14. Kim SH, Jeong JH, Kim TI, Kim SW, Bull DA (2009) VEGF siRNA delivery system using arginine-grafted bioreducible poly(disulfide amine). Mol Pharm 6:718–726

Chapter 11

Using LacO Arrays to Monitor DNA Double-Strand Break Dynamics in Live *Schizosaccharomyces pombe* Cells

Bryan A. Leland and Megan C. King

Abstract

LacO arrays, when combined with LacI-GFP, have been a valuable tool for studying nuclear architecture and chromatin dynamics. Here, we outline an experimental approach to employ the LacO/LacI-GFP system in *S. pombe* to assess DNA double-strand break (DSB) dynamics and the contribution of chromatin state to DSB repair. Previously, integration of long, highly repetitive LacO arrays in *S. pombe* has been a challenge. To address this problem, we have developed a novel approach, based on the principles used for homologous recombination-based genome engineering in higher eukaryotes, to integrate long, repetitive LacO arrays with targeting efficiencies as high as 70 %. Combining this facile LacO/LacI-GFP system with a site-specific, inducible DSB provides a means to monitor DSB dynamics at engineered sites within the genome.

Key words LacO/LacI, Chromatin dynamics, DNA double-strand break, Genome instability, Homologous recombination, Live-cell imaging, Genome engineering, *S. pombe*

1 Introduction

DNA damage is a major contributor to genome instability; disruption of many DNA repair processes can promote cellular transformation and oncogenesis [1]. Tremendous progress using biochemical and genetic approaches to identify and characterize the proteins and pathways used to repair DNA double-strand breaks (DSBs) has been made since the current model of DSB repair was first proposed [2]. However, because most studies to date have relied on population-based approaches like chromatin immunoprecipitation or Southern blots, there is still relatively little known about the dynamics of individual DSBs or the role that the nuclear environment plays in their repair. For example, a DSB must pair with an unbroken, homologous DNA strand to use as a template for repair. How this homolog pairing occurs is a topic of great interest, but this process has proven difficult to study [3–5].

Narendra Wajapeyee (ed.), *Cancer Genomics and Proteomics: Methods and Protocols*, Methods in Molecular Biology, vol. 1176, DOI 10.1007/978-1-4939-0992-6_11, © Springer Science+Business Media New York 2014

Much of the DSB repair literature takes advantage of the model eukaryote, *Saccharomyces cerevisiae*, to study DSB repair at the endogenous mating-type locus as well as several ectopically generated DSB sites [6]. As we learn more about DSB repair, it is becoming apparent that the chromatin context of a DSB impacts the mechanism used for its repair [7–12]. Thus, it is important to look beyond the *S. cerevisiae* mating-type locus and a few select ectopic sites and survey the mechanisms of DSB repair in a wide range of genomic contexts. In addition, *S. cerevisiae* lack many proteins involved in heterochromatin (HC) formation in higher eukaryotes, making it less well suited to study the role that HC plays in DSB repair [13].

To address these challenges, we have adapted a system for the visualization of a single, inducible DSB in live cells for use in *S. pombe*. Our method uses a novel, high-efficiency approach to integrate a LacO array, taking advantage of the principles used in homologous recombination-based genome engineering in higher eukaryotes [14]. After LacO array integration, a second round of integration is used to insert an adjacent, site-specific nuclease recognition site. This system allows for rapid assessment of live-cell dynamics of DSBs at many different genomic positions with varied chromatin states.

2 Materials

2.1 Plasmids and DNA

1. AscIcs-Ura-LacO-10.3kb plasmid (this publication, *see* Subheading 3.1): This plasmid will be used to generate large quantities of AscIcs-Ura-LacO plasmid with LacO array size(s) between 1.0 and 10.3 kb.
2. WT *S. pombe* genomic DNA (purified using the glass bead method [15] or another equivalent protocol).
3. Primers for amplifying megaprimers (*see* Table 1).
4. Primers for checking transformation candidates (*see* Table 1).
5. pFA6a-HOcs-HphR plasmid (this publication, *see* **Note 1**).
6. pREP81-HO plasmid [16] (*see* **Note 2**).

2.2 Molecular Biology Reagents

1. LB + Amp media.
2. LB + Amp agar plates.
3. MAX Efficiency Stbl2 Competent Cells.
4. Small-scale plasmid purification kit.
5. Medium- or large-scale plasmid purification kit.
6. AscI, HincII, and associated buffers.
7. 200 μL PCR tubes.
8. ddH_2O.

Table 1
Primer design

Name	Sequence	Genomic position	Strand
Left and right "megaprimers" to insert HOcs-HphR at future LacO site (Fig. 1a, b)			
MpL-F	5′-Genomic-3′	–300–500 bps	+
MpL-R	5′-GTCTGCTCCCGGCATCCGCT-Genomic-3′	–1–20 bps	–
MpR-F	5′-GGCGAGCGGTATCAGCTCAC-Genomic-3′	+1–20 bps	+
MpR-R	5′-Genomic-3′	+300–500 bps	–
Checking primers to verify proper HO-HphR targeting (Fig. 1c)			
Lchk-F	5′-Genomic-3′	–500–700 bps	+
Lchk-R	5′-GTATTCTGGGCCTCCATGTCGCTG-3′		–
Rchk-F	5′-CGCCTCGACATCATCTGCCCAGATG-3′		+
Rchk-F	5′-Genomic-3′	+500–700 bps	–
Checking primers to verify proper AscIcs-Ura-LacO targeting (Fig. 1e)			
Lchk-F	5′-Genomic-3′ (same as above)	–500–700 bps	+
Lchk-R	5′-TTATTGTCTCATGAGCGGATACAT-3′		–
Left and right "megaprimers" to insert HOcs-HphR adjacent to LacO (Fig. 1f)			
MpL-F	5′-Genomic-3′	–300–500 bps	+
MpL-R	5′-TTAATTAACCCGGGGATCCG-Genomic-3′	–1–20 bps	–
MpR-F	5′-GTTTAAACGAGCTCGAATTC-Genomic-3′	+1–20 bps	+
MpR-R	5′-Genomic-3′	+300–500 bps	–

First, primers are used to amplify an HOcs-HphR targeted cassette. Two of these primers (MpL-R and MpR-F) contain altF1/altR1 sequences that are necessary to amplify cassettes from pFA6a-derived plasmids that include specialized targeting sequences [18]. There will be 0–40 bps in the genome between MpL-R and MpR-F that will be deleted by this integration. Next, checking primers are used to verify proper targeting of the HOcs-HphR cassette, followed by successful replacement of HOcs-HphR with AscIcs-Ura-LacO. Finally, an HOcs-HphR targeted cassette that will be inserted adjacent to the LacO array is generated using the standard F2/R1 sequences [22]. All primers that include sequence from the *S. pombe* genome are indicated with "Genomic" in their sequence. These primers should be designed at sites in the *S. pombe* genome flanking the desired insertion location, as specified by their "Genomic position" and "Strand" (*see* Fig. 1). In general, primers should be designed to have 24–32 bps of homology with the *S. pombe* genome and a GC percentage of 30–70 %

9. Taq (or equivalent) DNA polymerase.
10. KOD (Millipore), iProof (Bio-Rad), or equivalent high-fidelity DNA polymerase.
11. LongAmp® Taq 2× Master Mix (NEB, for optional colony PCR reactions).
12. PCR purification kit.
13. 1 % Agarose gel and related supplies for electrophoresis/detection.

2.3 Yeast Growth and Selection

1. 250 mL Erlenmeyer flasks.
2. EMM5S media (all *S. pombe* media and reagents based on [17]).
3. EMM –Leu media.
4. YE5S plates.
5. YE5S +Hyg plates: Hygromycin B is added to a final concentration of 200 μg/mL after autoclaving.
6. EMM –Ura plates.
7. EMM –Leu –Ura plates.
8. EMM –Leu +Thi plates: Thiamine is added to a final concentration of 5 μg/mL before autoclaving. Add 500 μL of a 10 mg/mL stock solution per 1.0 L.
9. Sterile velvets (for replica plating).
10. Glass plate spreader.

2.4 Yeast Transformations

1. 50 mL Conical tubes.
2. 1.5 mL Microcentrifuge tubes.
3. Sterile ddH_2O.
4. Chilled LiAc-TE: 100 mM Lithium acetate, 10 mM Tris pH 7.5, 1 mM EDTA. Prepare immediately before use by combining 10× TE with 10× lithium acetate, and ddH_2O.
5. Carrier ssDNA from salmon testes, heat shock immediately before use.
6. LiAc-TE-PEG: 100 mM Lithium acetate, 10 mM Tris pH 7.5, 1 mM EDTA, 40 % PEG 4000. Prepare immediately before use from stocks of 10× TE, 10× lithium acetate, and 1.25× PEG.
7. DMSO, warmed to 37 °C.

2.5 Equipment

1. Thermocycler.
2. 42 °C Water bath.
3. 37 °C Incubator or water bath.
4. 30 °C Shaking incubator compatible with 250 mL flasks.
5. 30 °C Incubator.
6. Slow rotator for 1.5 mL microcentrifuge tubes (fits inside 30 °C incubator).
7. Centrifuge compatible with 50 mL conical tubes.
8. Microcentrifuge.
9. GFP-capable microscope: Any microscope with a 488 nm light source, accompanying GFP filters, and a reasonably sensitive camera may be used. We used a Deltavision Widefield Deconvolution Microscope (Applied Precision/GE Healthcare) with an Evolve 512 EMCCD camera (Photometrics).

3 Methods

3.1 Generate Sufficient Quantities of AscIcs-Ura-LacO Plasmid

1. We have made an AscIcs-Ura-LacO-10.3kb plasmid for integrating LacO arrays into *S. pombe*. This plasmid is derived from pSR10, which has 10.3 kb of LacO repeats (~256 copies of HaeIII-containing 36 bp LacO repeat units, arranged in ~32 copies of 316 bp 8-mer units that are separated by EcoRI) and a TRP1 marker for integration into *S. cerevisiae*; here TRP1 is replaced with URA4 for selection in *S. pombe* [18–21]. The AscIcs-Ura-LacO-10.3 kb plasmid and its full sequence are available upon request.
2. The repetitive nature of the 10.3 kb LacO array in the AscIcs-Ura-LacO-10.3kb plasmid makes it highly unstable in most bacteria. To maintain LacO array repeat number as much as possible, transform 1–50 ng of the AscIcs-Ura-LacO-10.3kb plasmid into MAX Efficiency Stbl2 Competent Cells (Invitrogen).
3. Follow the manufacturer's Stbl2 protocol exactly, including all incubations at 30 °C instead of 37 °C. Grow the candidate transformants on LB + Amp plates for 12–24 h at 30 °C.
4. Many of the candidates will have shortened LacO arrays. Screen 10–25 candidates for LacO array size by Mini-prep and HincII digest as follows:
   (a) Label 10–25 candidate colonies on the LB + Amp plate, wrap in parafilm, and store at 4 °C until the initial screening is completed, up to 1 week (*see* **Note 3**).
   (b) For each candidate, grow 1–5 mL cultures overnight in LB + Amp at 30 °C.
   (c) Recover plasmid DNA (Mini-prep, QIAGEN, or equivalent protocol).
   (d) Digest 200–500 ng of plasmid DNA from each candidate with HincII as directed. Run on a 1 % agarose gel. The HincII digest cuts out the LacO repeats from the AscIcs-Ura backbone. The backbone is 4.1 kb, and the LacO repeats will be between 1.0 and 10.3 kb and may have been dropped entirely in some candidates (*see* **Note 4**).
5. For all selected candidate(s), go back to the original colony (stored at 4 °C) within 1 week. Grow larger Midi- or Maxi-prep-sized LB + Amp culture(s) at 30 °C to obtain large quantities of properly sized LacO repeats.
6. Perform Midi- or Maxi-prep(s) as directed (QIAGEN or equivalent protocol) to obtain 50–1,000 μg of each AscIcs-Ura-LacO plasmid. Resuspend the plasmid(s) at ≥400 ng/μL; 3 μg of plasmid is required for each *S. pombe* transformation.

3.2 Generate a Targeted HOcs-HphR Cassette Using the "Two-Step" Method

1. Using WT genomic DNA as a template and Taq or any other standard DNA polymerase, PCR amplify the left and right "megaprimers" with standard buffers as directed for 35 cycles. *See* Fig. 1a and Table 1 for information about the positions and sequences of the four primers used for these two PCR reactions (*see* **Note 5**). The location of these primers in the genome dictates the insertion location of HOcs-HphR, which will subsequently be replaced by AscIcs-Ura-LacO (*see* **Note 6**). Importantly, the two outward-facing primers (MpL-R and MpR-F) must contain the altF2/altR1 sequences that will subsequently be used to amplify a pFA6a-derived plasmid [18, 22].

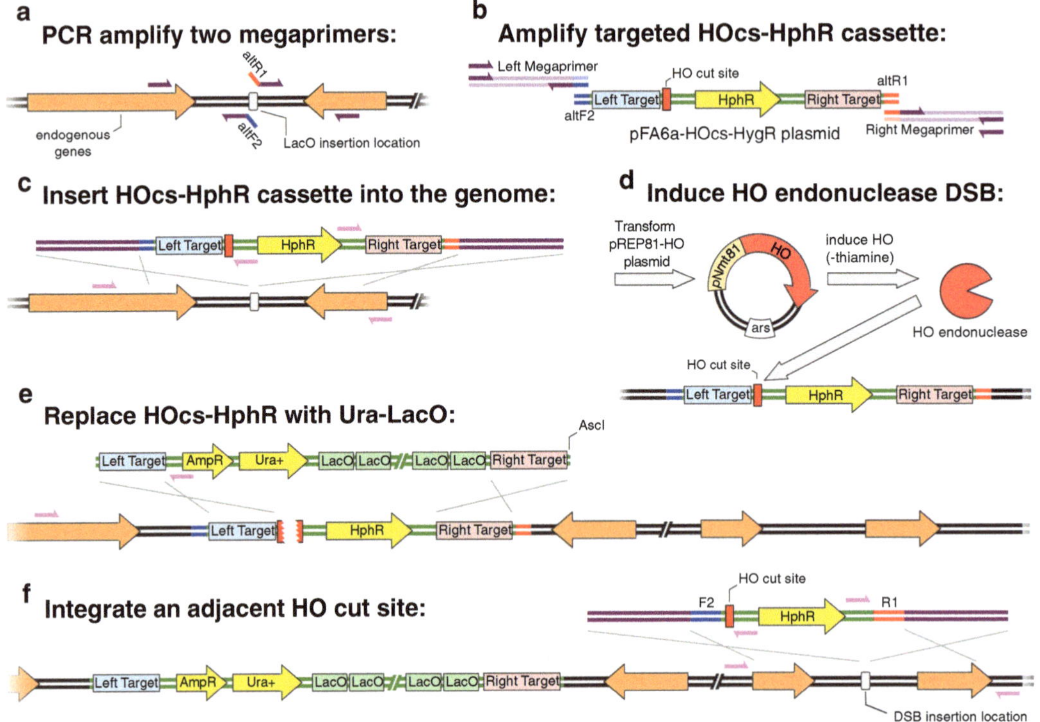

Fig. 1 LacO integration method. (**a**) Genomic DNA is used to amplify two "megaprimers" (*see* Table 1 for primer design). (**b**) The two megaprimers are used to amplify a targeted cassette from the pFA6a-HOcs-HphR plasmid. (**c**) The targeted HOcs-HphR cassette is integrated into the genome and checked for proper targeting with two pairs of primers shown in *pink* and described in Table 1. (**d**) A plasmid containing the HO endonuclease gene under the control of the Nmt81 promoter is introduced by transformation and initially repressed by the addition of thiamine. HO expression is then induced by withdrawal of thiamine, causing a site-specific DSB at the desired location of LacO integration. (**e**) Immediately following HO DSB induction, the AscI-linearized LacO array-containing plasmid is transformed and selection for integrants is carried out on EMM –Leu –Ura plates. When targeting is successful, the HphR marker will be replaced by the Ura marker and the HO cut site will be lost. (**f**) A second targeted cassette with HphR and an HO endonuclease cut site is integrated adjacent to the LacO array. This allows a single, site-specific DSB to be generated next to the LacO array

2. PCR purify (QIAGEN or equivalent kit) the left and right megaprimer products. Verify that the PCR reactions were successful and that the products are completely free of primers on a 1 % agarose gel (residual primers will disrupt the next amplification reaction).
3. Make a targeted HOcs-HphR cassette (*see* **Note 1**). Use KOD (Millipore), iProof (Bio-Rad), or another high-fidelity polymerase to PCR amplify the targeted HOcs-HphR cassette from the pFA6a-HOcs-HphR plasmid, as shown in Fig. 1b. Prepare a 50 μL PCR reaction as described for the specific polymerase, with these exceptions:
   (a) Add two primers to the reaction, MpL-F and MpR-R, both at final concentration of 0.5 μM (Table 1).
   (b) Add the PCR-purified left and right megaprimers, each to a final concentration of ~8 ng/μL (usually each 50 μL reaction will contain ~3 μL of each PCR-purified megaprimer).
   (c) Use 15–100 ng of pFA6a-HOcs-HphR as the template DNA.
   (d) Amplify for 35 cycles as directed. The product will be between 2.8 and 3.2 kb, depending on the length of the megaprimers.
4. Check the purity of the targeted HOcs-HphR cassette on a gel. Residual megaprimers are acceptable as long as the correct band for the targeted HOcs-HphR cassette is also clearly visible.

3.3 Integrate the Targeted HOcs-HphR Cassette into the Genome

1. Grow *S. pombe* cells overnight in 50 mL of EMM5S in a 250 mL Erlenmeyer flask at 30 °C with shaking at 230 rpm (*see* **Note 7**). The strain used must be HphR minus and a uracil and leucine double auxotroph (*leu1-32*; *ura4-D18*). It should also already have a stably integrated LacI-GFP, or this must be subsequently crossed into the strain (*see* **Note 8**).
2. The next day, harvest the cells when they reach an OD_{600} of between 0.5 and 1.0. Centrifuge at 1,000 × *g* for 10 min in a 50 mL conical tube. If the culture has already reached saturation in the morning, it may be diluted and grown for another 2–6 h until it returns to OD_{600} = 0.5–1.0.
3. Discard the supernatant, resuspend the cell pellet in 1 mL of sterile ddH_2O, and transfer to a 1.5 mL microcentrifuge tube.
4. Wash by spinning for 1 min at 1,000 × *g* to pellet the cells, and then discard the supernatant. Wash 2× with 1 mL sterile ddH_2O and then 2× with 1 mL chilled, freshly prepared LiAc-TE.
5. Estimate the volume of the wet cell pellet and resuspend in about 2 volumes of chilled LiAc-TE. Resuspended cells may be transformed immediately or stored on ice for up to 4 h.

6. Heat shock carrier ssDNA at 95 °C for 5 min and then immediately transfer to an ice bath for 5 min.
7. Combine 10 μL carrier ssDNA, 2.5 μL targeted HphR cassette (directly from the PCR reaction, no purification necessary), and 50 μL of cells resuspended in LiAc-TE. Incubate at RT for 10 min.
8. Add 260 μL of freshly prepared LiAc-TE-PEG, and vortex. Incubate at 30 °C for 1 h on a slow rotator.
9. Add 43 μL of DMSO, pre-warmed to 37 °C. Heat shock the cells for 5 min in a 42 °C water bath.
10. Pellet the cells by spinning for 1 min at 1,000 × *g*. Discard the supernatant. Resuspend the pellet in 150 μL sterile ddH_2O.
11. Spread all 150 μL of transformed cells onto a YE5S plate with a sterile glass spreader.
12. Grow overnight for 14–16 h at 30 °C. The following morning, use a sterile velvet to replica plate to a YE5S +Hyg plate for selection.
13. Screen for correctly targeted, HphR-positive candidates using checking PCR primers (*see* **Note 9**). Checking primers are shown in pink in Fig. 1c, and sequences are listed in Table 1.

3.4 Transform an HO Endonuclease Episomal Plasmid

1. Using HO endonuclease to make a site-specific DSB at the insertion location before transforming the LacO array increases targeting efficiency (Fig. 1d, e). It is possible to skip this step and integrate the LacO array without first generating an HO DSB, but the targeting efficiency is so low (0.1–2 %) that this is not recommended. In contrast, targeting efficiencies after generating an HO DSB can be as high as 70 %.
2. Transform the pREP81-HO plasmid with the same transformation procedure as above (Subheading 3.3) with the following modifications:
   (a) Subheading 3.3, **step** 7: Combine 10 μL carrier ssDNA, ~200 ng of pREP81-HO plasmid (*see* **Note 2**), and 50 μL of cells resuspended in LiAc-TE. Incubate at RT for 10 min.
   (b) Subheading 3.3, **step 11**: Plate the cells directly to EMM –Leu +Thi plates (*see* **Note 10**). Depending on the amount of plasmid used, it may be necessary to plate ≤90 μL of the resuspended cells to obtain single colonies. Grow at 30 °C for 4–8 days.

3.5 Transform Linearized AscIcs-Ura-LacO (Highly Concentrated Transformation)

1. Induce HO DSB formation, and then immediately transform 3 μg of digested AscIcs-Ura-LacO (*see* **Note 11** and Fig. 1d, e). Use the same transformation procedure as above (Subheading 3.3), with the following exceptions:

(a) Subheading 3.3, **step 1**: Grow cells overnight in 100 mL of EMM –Leu in a 250 mL Erlenmeyer flask at 30 °C with shaking at 230 rpm. This media will induce the expression of the HO endonuclease due to the lack of thiamine.

(b) For each transformation, linearize 3 μg of AscIcs-Ura-LacO. Digest in a 10 μL reaction 3 μg of AscIcs-Ura-LacO, 1× NEB4, and 20 U of AscI (NEB). Perform the reaction at 37 °C for 4–5 h for complete digestion.

(c) Subheading 3.3, **step 5**: Estimate the volume of the wet cell pellet, and make a highly concentrated resuspension by adding about 0.75–1 volumes of chilled LiAc-TE.

(d) Subheading 3.3, **step 7**: Combine all 10 μL (=3 μg) of the digestion reaction with 10 μL ssDNA (heat shocked) and 50 μL resuspended cells.

(e) Subheading 3.3, **step 11**: Plate cells directly, spreading them across 2–3 EMM –Leu –Ura plates. It is advisable to spread the 150 μL of resuspended cells across multiple plates (20–90 μL per plate) to obtain many single colonies. Grow at 30 °C for 4–6 days.

3.6 Screening Candidates (Ura Positive, HphR Minus)

1. After 4–6 days, pick single colonies to screen by patching them to YE5S plates (*see* **Note 12**). Grow patched candidates at 30 °C for 1–3 days.
2. After patched candidates have completely grown up on the YE5S plates, use a sterile velvet to replica plate to YE5S +Hyg, EMM –Ura, and EMM –Leu plates. Grow plates at 30 °C for 1–3 days.
3. Candidates that grow on EMM5S –Ura but die on YE5S +Hyg have AscIcs-Ura-LacO successfully inserted at the correct location.
4. Some candidates with correctly targeted AscIcs-Ura-LacO may also have already lost the pREP81-HO plasmid (they will die on EMM –Leu). If not, pREP81-HO should be removed before subsequent transformations or experiments (*see* **Note 13**).

3.7 Additional Candidate Screening Methods

1. Screening by PCR: The Ura-positive, HphR-minus candidates can also be double-checked by PCR across the left target (*see* **Note 9**). Primers for this checking reaction are shown in pink in Fig. 1e, and sequences are given in Table 1.
2. Screening by microscopy: If a strain background with LacI-GFP was used (*see* **Note 8**), the LacO/LacI-GFP focus should be immediately visible on a microscope. Any standard bright-field or confocal microscope with a moderately strong 488 nm excitation source and appropriate emission filter may be used (Fig. 2).

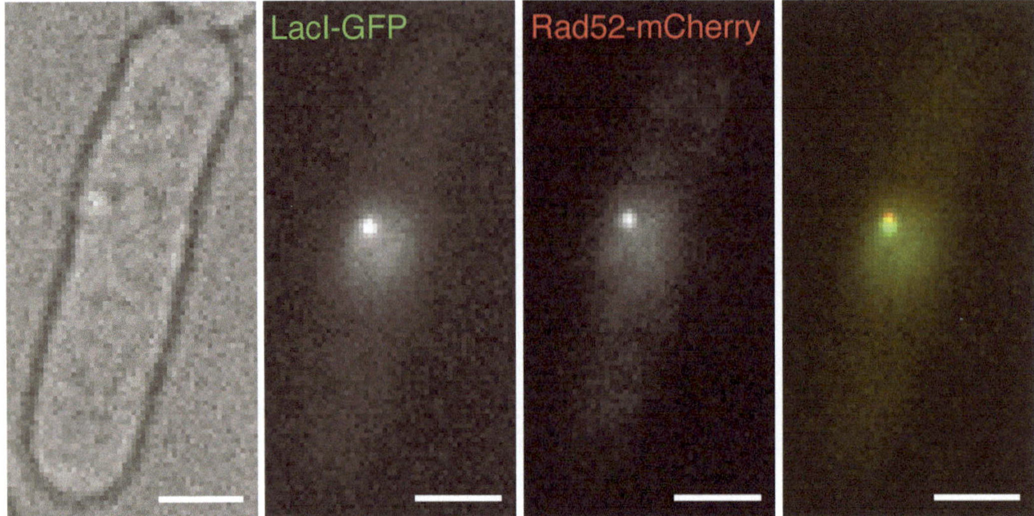

Fig. 2 Example fluorescence micrographs of a LacO/LacI-GFP labeling an adjacent HO DSB. A site-specific DSB was generated by transforming pREP81-HO and inducing HO expression in EMM –Leu –Thi media for 24 h. Rad52 is a DSB-binding protein involved in repair. Images are a single Z section through the middle of the cell. Scale bars = 3 μm

(a) Log-phase cells (OD_{600} = 0.5–1.0) should be viewed on the microscope because saturated cells have variable LacI-GFP expression.

(b) Check for cell health/viability when screening candidates on the microscope (*see* **Note 6**). Different candidates from the same LacO transformation may have slightly variable LacO/LacI-GFP focus intensity, so they should be compared to select an optimal strain (*see* **Note 4**).

3.8 Transforming an HO Cut Site Adjacent to the LacO Array

1. Ensure that the pREP81-HO episomal plasmid has been dropped before beginning this transformation (*see* Subheading 3.6, **step 4**).
2. Using the same pFA6a-HOcs-HphR plasmid, generate a targeted HOcs-HphR cassette to integrate adjacent to the LacO array (Fig. 1f, Table 1). The targeted cassette is generated in the same way as described above in Subheading 3.2, although F2/R1 sequences are now used instead of altF2/altR1 (*see* **Note 14**).
3. Perform and validate the transformation of HOcs-HphR as described above in Subheading 3.3. Checking primers are also designed in the same way, as described in Table 1.

3.9 Express HO Endonuclease to Generate a LacO-Proximal DSB

1. Transform the pREP81-HO plasmid into the new strain containing HOcs-HphR adjacent to AscIcs-Ura-LacO as well as LacI-GFP and optionally Rad52-mCherry or another early DSB marker (*see* **Note 15**).

2. Perform the transformation as above (Subheading 3.4) with ~200 ng of the pREP81-HO episomal plasmid. Be sure to plate directly to EMM –Leu +Thi plates after the transformation to keep the HO endonuclease repressed.
3. To induce HO expression and DSB formation, grow cells overnight in EMM –Leu (no thiamine, induced). Induction is most reliable in the first 1–2 weeks after transformation, so it is best to use pREP81-HO transformants as soon as possible.
4. DSB formation occurs between 16 and 24 h, approximately when cells reach log phase (OD_{600} = 0.5–1.0). DSB formation and timing can be monitored by the appearance of Rad52-mCherry foci that colocalize with LacO/LacI-GFP (Fig. 2).
5. LacO/LacI-GFP dynamics can be assessed by acquiring 5–40 Z slices through the nucleus (~3 μm) over a 0.5–2.5-h time course. The nuclear envelope can be roughly determined by the diffuse nuclear LacI-GFP that is not bound to the LacO array or, alternatively, a nuclear envelope marker may be used (e.g.: Nup107-mCherry).

4 Notes

1. We have created a pFA6a-HOcs-HphR plasmid that can be used to generate HOcs-HphR targeted cassettes. The 35-bp HO endonuclease cut site was inserted between the F2 sequence and the HphR gene of the pFA6a-HphR plasmid [6, 22, 23]: HO cut site: AGTTTCAGCTTTCCGC**AACA**G TATAATTTTATAAA.
2. The pREP81-HO plasmid [16] contains Leu as a selectable marker and the HO endonuclease coding sequence under the inducible expression of the pNmt81 promoter [24].
3. The LacO array length will be stable on a plate at 4 °C for 1 week, but going back to frozen stocks of bacteria is not recommended as array number can decrease. Instead, to obtain more plasmid, it is necessary to re-transform a plasmid containing the longest available LacO repeat (10.3 kb) into Stbl2 cells again, and then screen for the desired length of LacO repeats by HincII digest.
4. It may be beneficial to select several candidates with LacO array sizes that range from 1.0 to 10.3 kb for further purification. 2.5–10 kb LacO arrays are easier to detect by microscopy with LacI-GFP, although the full 10.3 kb is generally not necessary for moderately sensitive microscope setups. It is important to note that longer LacO arrays inserted into the *S. pombe* genome are more likely to induce heterochromatization, affect nuclear localization, and/or cause viability defects [25, 26].

5. The protocol described here is adapted from the "two-step" method developed by the lab of Michael Keogh. It is also possible to generate an HphR cassette with a single-step PCR method that uses long primers with 60–80 bp of homology to the genome [22]. The two-step method uses "megaprimers" to generate integration cassettes with 300–500 bp of homology. We have found that these longer targeting sequences increase the targeting efficiency, especially in AT-rich or repetitive intergenic regions; however both methods are acceptable.
6. Inserting repetitive LacO arrays (especially those >2.5 kb) at some places in the genome can impact growth/viability, even when the insertion location is intergenic. It is advisable to try 2–4 candidate loci to ensure that at least one gives efficient LacO integration and has a minimal impact on growth/viability.
7. Alternatively, YE5S or other media may be used if required for growth of the desired strain.
8. Strains derived from Nabeshima et al. expressing LacI-GFP under the pDis1 promoter are widely available [27, 28]. Note that there is no auxotrophic or resistance-based marker on the LacI-GFP in these strains. However, LacI-GFP is linked to the mating-type locus, so it will segregate with mating type in crosses.
9. It is possible to screen candidates by purifying genomic DNA [15] or by performing colony PCR. For colony PCR, take a small clump of cells (0.1–0.3 mg, just visible on a small pipette tip) directly from a fresh plate (<5 days old). Mix the cells into a PCR reaction containing 10 μL of LongAmp Taq 2× Master Mix, 1.25 μM of each checking primer, and H_2O up to 20 μL total volume. We have found that while many polymerases fail in colony PCR reactions, the 2× LongAmp Taq Master Mix is very robust. Precede cycling with a 95 °C, 13-min incubation to lyse the cells. Then run 30 cycles of 95 °C, 30 s; 45–65 °C, 30 s; and 65 °C, 90 s/kb.
10. Replica plating is not necessary for this Leu selection. Adding thiamine to the plate is essential; this keeps HO endonuclease repressed.
11. A large amount of linearized AscIcs-Ura-LacO (3 μg) is necessary for this transformation to increase the efficiency as much as possible. This transformation relies on recombination between the "left target" and "right target" sequences (*see* Fig. 1e). These sequences are only present in the genome when the targeted cassette is generated using altF2/altR1 primer sites (Subheading 3.2). Therefore, strain backgrounds harboring an existing pFA6a-based integration can still be used, provided they were generated with the standard F2/R1 primer sites [22].

12. Wait until even the smaller colonies on the plate are sufficiently large to pick and restreak. Sometimes LacO integrations can reduce growth, in which case the desired, properly targeted strains will be the smaller colonies. If you skipped Subheading 3.4 above (transformation of pREP81-HO episomal plasmid), then you will need to screen between 100 and 500 candidates to find one with a properly targeted LacO array. If you generated an HO DSB at the insertion location first, then the LacO targeting should be much more efficient; screening 10–50 candidates is sufficient.
13. The pREP81-HO episomal plasmid is retained by –Ura –Leu selection immediately after the transformation because this increases the AscIcs-Ura-LacO targeting efficiency. Before subsequent transformations (Subheading 3.8) or experiments involving HO cut sites, this plasmid should be removed. Loss of pREP81-HO can be achieved by backcrossing or simply by removal of –Leu selection for several generations due to the high loss rate of episomal plasmids in *S. pombe* [29, 30]. Loss may occur naturally during subsequent screening (Subheading 3.7). If not, AscIcs-Ura-LacO integrants can be struck to a YE5S plate for single colonies. Individual colonies can then be screened for the loss of pREP81-HO on EMM –Leu plates; the loss rate is >50 %.
14. After the HO enzyme cuts at its recognition site, resection begins on either side of this DSB and can continue for many kb because the DSB cannot be repaired (both sister chromatids are cut). Therefore, for longer time course experiments it is necessary to integrate the HOcs-HphR >20 kb away from the LacO array. Also, since HO induction from the pREP81-HO plasmid is asynchronous, it may be beneficial to have a marker of DSB formation in the strain background (e.g.: Rad52-mCherry).
15. There are several different systems for inducing HO DSBs in *S. pombe* [31, 32]. Here we describe how to generate a DSB adjacent to the LacO array using the same pREP81-HO plasmid described above in Subheading 3.4 (*see* **Note 2**). Alternatively, a plasmid that generates higher levels of DSBs, pREP41-HO, is also available.

Acknowledgements

We would like to thank the Gasser lab and Russell labs for providing plasmids. This work was supported by the G. Harold and Leila Y. Mathers Charitable Foundation and the Searle Scholar Program (to M.C.K) and an NIGMS training grant T32GM007223 (to B.A.L.).

References

1. Negrini S, Gorgoulis VG, Halazonetis TD (2010) Genomic instability–an evolving hallmark of cancer. Nat Rev Mol Cell Biol 11:220–228
2. Szostak JW, Orr-Weaver TL, Rothstein RJ et al (1983) The double-strand-break repair model for recombination. Cell 33:25–35
3. Barzel A, Kupiec M (2008) Finding a match: how do homologous sequences get together for recombination? Nat Rev Genet 9:27–37
4. Gehlen LR, Gasser SM, Dion V (2011) How broken DNA finds its template for repair: a computational approach. Prog Theor Phys Suppl 191:20–29
5. Dion V, Gasser SM (2013) Chromatin movement in the maintenance of genome stability. Cell 152:1355–1364
6. Haber JE (2012) Mating-type genes and MAT switching in Saccharomyces cerevisiae. Genetics 191:33–64
7. Goodarzi AA, Noon AT, Deckbar D et al (2008) ATM signaling facilitates repair of DNA double-strand breaks associated with heterochromatin. Mol Cell 31:167–177
8. Chiolo I, Minoda A, Colmenares SU et al (2011) Double-strand breaks in heterochromatin move outside of a dynamic HP1a domain to complete recombinational repair. Cell 144: 732–744
9. van Attikum H, Fritsch O, Hohn B et al (2004) Recruitment of the INO80 complex by H2A phosphorylation links ATP-dependent chromatin remodeling with DNA double-strand break repair. Cell 119:777–788
10. Costelloe T, Louge R, Tomimatsu N et al (2012) The yeast Fun30 and human SMARCAD1 chromatin remodellers promote DNA end resection. Nature 489:581–584
11. Nagai S, Dubrana K, Tsai-Pflugfelder M et al (2008) Functional targeting of DNA damage to a nuclear pore-associated SUMO-dependent ubiquitin ligase. Science 322: 597–602
12. Chen X, Cui D, Papusha A et al (2012) The Fun30 nucleosome remodeller promotes resection of DNA double-strand break ends. Nature 489:576–580
13. Grewal SIS, Jia S (2007) Heterochromatin revisited. Nat Rev Genet 8:35–46
14. van der Oost J (2013) Molecular biology. New tool for genome surgery. Science 339: 768–770
15. Rose MD, Winston FM, Heiter P (1990) Methods in yeast genetics: a laboratory course manual, Cold Spring Harbor Laboratory Protocols. Cold Spring Harbor, NY
16. Du L-L, Nakamura TM, Moser BA et al (2003) Retention but not recruitment of Crb2 at double-strand breaks requires Rad1 and Rad3 complexes. Mol Cell Biol 23:6150–6158
17. Moreno S, Klar A, Nurse P (1991) Molecular genetic analysis of fission yeast Schizosaccharomyces pombe. Methods Enzymol 194: 795–823
18. Rohner S, Gasser SM, Meister P (2008) Modules for cloning-free chromatin tagging in Saccharomyces cerevisiae. Yeast 25:235–239
19. Straight AF, Belmont AS, Robinett CC et al (1996) GFP tagging of budding yeast chromosomes reveals that protein-protein interactions can mediate sister chromatid cohesion. Curr Biol 6:1599–1608
20. Robinett CC, Straight A, Li G et al (1996) In vivo localization of DNA sequences and visualization of large-scale chromatin organization using lac operator/repressor recognition. J Cell Biol 135:1685–1700
21. Belmont AS (2001) Visualizing chromosome dynamics with GFP. Trends Cell Biol 11: 250–257
22. Bähler J, Wu JQ, Longtine MS et al (1998) Heterologous modules for efficient and versatile PCR-based gene targeting in Schizosaccharomyces pombe. Yeast 14:943–951
23. Pâques F, Haber JE (1997) Two pathways for removal of nonhomologous DNA ends during double-strand break repair in Saccharomyces cerevisiae. Mol Cell Biol 17:6765–6771
24. Basi G, Schmid E, Maundrell K (1993) TATA box mutations in the Schizosaccharomyces pombe nmtl promoter affect transcription efficiency but not the transcription start point or thiamine repressibility. Gene 123: 131–136
25. Jovtchev G, Watanabe K, Pecinka A et al (2008) Size and number of tandem repeat arrays can determine somatic homologous pairing of transgene loci mediated by epigenetic modifications in Arabidopsis thaliana nuclei. Chromosoma 117:267–276
26. Towbin BD, Meister P, Pike BL et al (2010) Repetitive transgenes in C. elegans accumulate heterochromatic marks and are sequestered at the nuclear envelope in a copy-number- and lamin-dependent manner. Cold Spring Harb Symp Quant Biol 75:555–565
27. Nabeshima K, Kurooka H, Takeuchi M et al (1995) p93dis1, which is required for sister chromatid separation, is a novel microtubule and spindle pole body-associating protein

phosphorylated at the Cdc2 target sites. Genes Dev 9:1572–1585
28. Nabeshima K, Nakagawa T, Straight AF et al (1998) Dynamics of centromeres during metaphase-anaphase transition in fission yeast: Dis1 is implicated in force balance in metaphase bipolar spindle. Mol Biol Cell 9: 3211–3225
29. Hayles J, Nurse P (1992) Genetics of the fission yeast Schizosaccharomyces pombe. Annu Rev Genet 26:373–402
30. Siam R, Dolan WP, Forsburg SL (2004) Choosing and using Schizosaccharomyces pombe plasmids. Methods 33:189–198
31. Sunder S, Greeson-Lott NT, Runge KW et al (2012) A new method to efficiently induce a site-specific double-strand break in the fission yeast Schizosaccharomyces pombe. Yeast 29:275–291
32. Watson AT, Werler P, Carr AM (2011) Regulation of gene expression at the fission yeast Schizosaccharomyces pombe urg1 locus. Gene 484:75–85

Chapter 12

Zebrafish as a Platform to Study Tumor Progression

Corrie A. Painter and Craig J. Ceol

Abstract

The zebrafish has emerged as a powerful model system to study human diseases, including a variety of neoplasms. Principal components that have contributed to the rise in use of this vertebrate model system are its high fecundity, ease of genetic manipulation, and low cost of maintenance. Vital imaging of the zebrafish is possible from the transparent embryonic stage through adulthood, the latter enabled by a number of mutant lines that ablate pigmentation. As a result, high-resolution analyses of tumor progression can be accomplished in vivo. Straightforward transgenesis of zebrafish has been employed to develop numerous tumor models that recapitulate many aspects of human neoplastic disease, both in terms of pathologic and molecular conservation. The small size of zebrafish embryos has enabled screens for novel chemotherapeutic agents. Its facile genetics have been exploited in studies that extend beyond modeling cancer to investigations that define new cancer genes and mechanisms of cancer progression. Together, these attributes have established the zebrafish as a robust and versatile model system for investigating cancer. In this chapter we describe methods that are used to study a gene's impact on melanoma progression. We detail methods for making transgenic animals and screening for tumor onset as well as methods to investigate tumor invasion and propagation.

Key words *Danio rerio*, Zebrafish, Melanoma, Melanocyte, BRAF, p53, Tumor progression

1 Introduction

The zebrafish has developed a solid foothold in cancer research. A variety of models have been generated due to the fact that most tissues in the zebrafish can become neoplastic [1, 2]. In addition to similarities between human and zebrafish tissues, there are a number of attributes that have contributed to the rise in popularity of this organism for cancer research, namely, its transparency, high fecundity, tractable genetics, and small size. These attributes allow for cellular resolution of tumor progression in intact animals, high statistical confidence in experimentation, the ability to investigate genes in autochthonous tumor models, and the ability to screen drugs in vivo. Together, these properties have paved the way for innovative studies of tumor progression and treatment.

Narendra Wajapeyee (ed.), *Cancer Genomics and Proteomics: Methods and Protocols*, Methods in Molecular Biology, vol. 1176, DOI 10.1007/978-1-4939-0992-6_12, © Springer Science+Business Media New York 2014

Cancer research in zebrafish has experienced rapid progress. Just 10 years ago it was first reported that tumors can be reliably induced by transgenesis [3]. Since then several tumor models have been generated including models targeting the skin, liver, intestine, pancreas, germ cells, and central nervous system [4–10]. Mesenchymal derived neoplasms have also been described including hemangiosarcoma, liposarcoma, and rhabdomyosarcoma [11–13]. These studies highlight the ease with which cancer models can be generated in the zebrafish.

Studies of cancer in the zebrafish have benefitted from technical advances and the availability of genomic information such that sophisticated strategies are now being used to unravel the genetic and mechanistic underpinnings of tumorigenesis. Improvements in transgenesis, primarily the incorporation of transposon-mediated integration [14], have facilitated the construction of stable transgenic strains. In addition, the high efficiency of transgenesis enables certain analyses to be conducted in chimeric transgenic animals, which forgoes the need to establish transgenic strains for some investigations. Fluorescent reporter transgenes can be easily utilized for imaging cell populations of interest in intact animals, as well as for cell sorting, obviating the need for technically challenging antibody-based labeling strategies. The advent of chemical screening in the zebrafish has opened the window to screens for compounds that affect tumorigenesis [15, 16]. Lastly, large-scale genetic screens can be undertaken to uncover novel or suspected links to tumorigenesis.

Below we present methods for assessing candidate tumor-promoting genes in an autochthonous melanoma model in zebrafish. Evaluation of candidate genes is performed in chimeric animals so that stable lines for each gene need not be generated. This feature increases the throughput of the approach as well as reduces the effort and cost associated with making stably transgenic lines. Chimeric animals are readily identified, and the transition of transgenic melanocytes to malignant melanomas is readily apparent based on morphological transitions. The melanomas that arise express the candidate gene in question, so studies of gene function requiring tumor sectioning, nucleic acid isolation, and cell sorting can be carried out with precision. The strategy outlined pertains to melanoma biology, although in principle neoplasms of other tissues can be studied by harnessing many of the same features of this model.

2 Materials

2.1 Gel Electrophoresis

1. 0.8 % Agarose (Fisher) in Tris base, acetic acid, EDTA (TAE) buffer.
2. Electrophoresis chamber (Bio-Rad).
3. Ethidium bromide (Pierce).
4. Gel extraction kit (Qiagen).

2.2 Construction of miniCoopR Clones

1. Herculase II Fusion DNA Polymerase (Stratagene).
2. Multisite Gateway Three-Fragment Vector Construction Kit (Invitrogen).

   pDONRP4-P1R vector (Invitrogen).

   pDONRP2R-P3 vector (Invitrogen).

   pDONRP2R-P3 vector (Invitrogen).
3. BP Clonase II Enzyme Mix (Invitrogen).
4. LR Clonase II Plus Enzyme Mix (Invitrogen).
5. 0.5 μl 10 mg/ml Proteinase K (Invitrogen).
6. Chemically competent DH5α *E. coli.*
7. LB broth (Fisher Scientific).
8. Agar (Fisher Scientific).
9. Kanamycin (Life Technologies).
10. Ampicillin (Sigma).
11. One Shot TOP10 Chemically Competent *E. coli* (Invitrogen).

2.3 Making and Selecting Transgenic Animals

1. Tol2 transposase mRNA (clone and protocols available via http://chien.neuro.utah.edu).
2. Borosilicate capillary (Sutter BF100-50-10 with filament).
3. Disposable scalpel (Feather Safety Razor Co., LTD).
4. Mineral oil (Sigma).
5. Stage micrometer with increments of 0.01 mm or less (Fisher Scientific).
6. Micromanipulator (Tritech Research, model GJ-1).
7. Injection apparatus (Harvard Apparatus, model PLI-100).
8. Dissection stereomicroscope (Leica, model M80).
9. Petri dish (Corning #430591).

2.4 Screening for Melanoma Onset and Studies of Tumor Tissue

1. 4.0 % Paraformaldehyde solution (Sigma).
2. 500 mM EDTA solution to decalcify (Sigma).
3. Trizol (Life Technologies).
4. Dissection scissors (Fine Science Tools).
5. 0.9× PBS supplemented (Corning).
6. 5 % FBS (Invitrogen).

3 Methods

The method we use to analyze the ability of a gene to modulate melanoma progression is outlined in Fig. 1. The tumor model on which it is based combines *Tg(mitfa:BRAF*^{V600E}), a melanocyte-expressed,

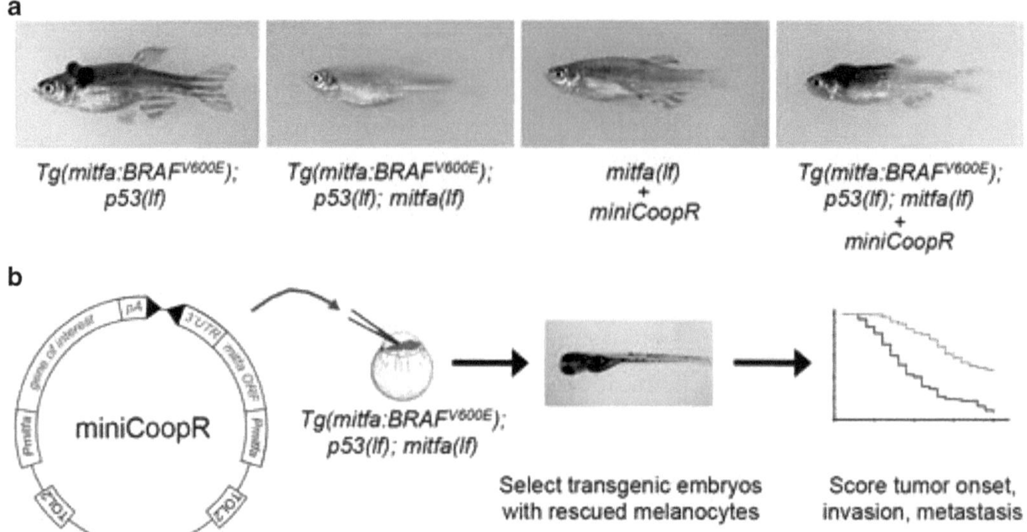

Fig. 1 Expression of transgenes in melanocytes and melanomas using the miniCoopR technique. (**a**) *Left*, a *Tg*(*mitfa*:*BRAF*^{V600E}); *p53*(*lf*) mutant with a dorsal melanoma. *Left center*, a *Tg*(*mitfa*:*BRAF*^{V600E}); *p53*(*lf*); *mitfa*(*lf*) mutant lacking melanocytes and melanoma. *Right center*, a *mitfa*(*lf*) mutant that was injected at the single-cell stage with a rescuing *mitfa* transgene. The transgene integrates into the genome mosaically, creating a chimeric animal with an incomplete pigment pattern. *Right*, the *mitfa* minigene rescues melanocyte development and tumor formation when injected into *Tg*(*mitfa*:*BRAF*^{V600E}); *p53*(*lf*); *mitfa*(*lf*) mutants. (**b**) Using the miniCoopR vector to express candidate oncogenes in melanomas. A miniCoopR-GOI clone is injected into single-cell *Tg*(*mitfa*:*BRAF*^{V600E}); *p53*(*lf*); *mitfa*(*lf*) embryos, and melanocyte-rescued animals are selected at 3 dpf. Rescued melanocytes also express the candidate oncogene because it is physically coupled to the *mitfa* minigene. Ultimately some melanocytes develop into melanomas having expressed the candidate oncogene throughout. Melanoma onset is monitored and tumor tissue harvested for analysis

stably integrated *BRAF*^{V600E} transgene, with a loss-of-function mutation in the zebrafish *p53* gene [4]. *Tg*(*mitfa:BRAF*^{V600E}); *p53*(*lf*) animals develop melanomas that are externally visible. The method also utilizes mutants that are defective in the *mitfa* gene, which acts cell autonomously to specify the melanocyte fate [17]. *mitfa* strains, including those in which the *mitfa:BRAF*^{V600E} transgene and *p53* mutation are present, lack melanocytes (Fig. 1a). This defect can be rescued by a *mitfa* minigene, which is made up of the *mitfa* promoter, *mitfa* open reading frame, and *mitfa* 3′ UTR. When the *mitfa* minigene is injected to make transgenic animals, the presence of melanocytes on an otherwise unpigmented background easily identifies transgene-positive animals and transgene-positive cells in those mosaic animals. If physically coupled to the *mitfa* minigene, a second transgene can be expressed in rescued melanocytes. In this way melanocytes expressing a transgene of choice can be easily monitored, and since the assay is performed in mosaics, many such transgenes can be assessed. To ease the DNA cloning required for transgenic studies, a Gateway recombination-compatible vector that contains the *mitfa* minigene is used (Fig. 1b). It is named miniCoopR for *mini*gene *coop*erating transgene *r*ecombination vector. miniCoopR utilizes multisite Gateway technology, in which pieces of DNA from three different plasmids can be introduced into a fourth plasmid via a single recombination reaction. Using this technology, cloning becomes modular, and many different promoter elements, open reading frames, and 3′ elements can be assembled with each other freely. Below are protocols describing the construction of miniCoopR clones containing genes of interest (GOIs), injection of miniCoopR-GOI clones, selection of transgene-positive animals, and screening for melanoma onset and progression.

3.1 Production of miniCoopR-GOI Clones for Melanoma Screening

miniCoopR-GOI clones are prepared using the Multisite Gateway Three-Fragment Vector Construction Kit (Invitrogen) (*see* **Note 1**).

3.1.1 Generation of the 5′ Promoter Clone

Promoter DNA generated by polymerase chain reaction (PCR) amplification is recombined into the pDONRP4-P1R vector (Invitrogen). Primers for amplification contain 20–25 nucleotides of promoter-specific sequence. In most miniCoopR-GOI clones, a 2.1 kilobase fragment of the *mitfa* promoter is used to drive melanocyte-specific expression (*see* **Note 2**).

1. Amplify the promoter from zebrafish genomic DNA using a high-fidelity polymerase and manufacturer-recommended PCR conditions. A typical reaction using Herculase II Polymerase is set up as follows:

1 μl	Template DNA (genomic DNA at 1 μg/μl, plasmid at 10 ng/μl)
1.5 μl	Forward primer (20 pmol/μl)
1.5 μl	Reverse primer (20 pmol/μl)
5 μl	dNTPs (2 mM each)
10 μl	10× Herculase II reaction buffer
0.5 μl	Herculase II Polymerase
30.5 μl	ddH_2O
50 μl	Total

2. Electrophorese the PCR reaction on an agarose gel to separate the promoter DNA from primer and unwanted PCR products.
3. Cut out the amplified promoter DNA using a low-intensity ultraviolet transilluminator, and purify using a gel extraction kit (Qiagen).

4. Insert purified promoter element into the entry clone through a BP reaction by mixing 25 fmol of purified promoter DNA together with 75 ng pDONRP4-P1R vector. Add nuclease-free water to 4 μl.
5. After thawing and vortexing BP Clonase II Enzyme Mix (Invitrogen), add 1 μl to the reaction and mix.
6. Incubate reaction mixture at 25 °C for >1 h.
7. Add 0.5 μl 10 mg/ml Proteinase K (Invitrogen), and incubate at 37 °C for 10 min.
8. Transform 2 μl of the reaction mix into 25 μl of DH5α *E. coli* (NEB). Follow standard procedures for transformation.
9. Spread 150 μl of transformation mix onto LB agar plates containing 50 μg/ml kanamycin.
10. Incubate overnight at 37 °C, and screen two or more resulting colonies for the correct, recombined promoter entry clone.

3.1.2 Generation of the Middle Open Reading Frame Clone

The gene of interest being studied is amplified by PCR and recombined into the pDONR221 vector (Invitrogen) (*see* **Note 3**).

1. The middle open reading frame clone is constructed using the same BP Clonase II protocol as above, except that a full-length open reading frame clone is used as a template for PCR amplification.

3.1.3 Generation of the 3′ Element Clone

3′ Elements are PCR amplified and recombined into the pDONRP2R-P3 vector (Invitrogen) (*see* **Note 4**).

1. The 3′ element clone is constructed using the same protocol as above using a clone containing the desired 3′ element as a template for PCR amplification. In most experiments, a 3′ element clone containing the SV40 polyadenylation sequence is suitable. Such clones have been constructed and are available through several sources [18, 19].

3.2 Multisite Gateway Cloning into miniCoopR Vector

1. Set up a multisite Gateway reaction by mixing 10 fmol of each entry clone (5′ promoter, middle open reading frame, and 3′ element) together with 10 fmol of miniCoopR vector. Add nuclease-free water to 4 μl.
2. After thawing and vortexing LR Clonase II Plus Enzyme Mix (Invitrogen), add 1 μl to the reaction and mix.
3. Incubate at 25 °C for >10 h.
4. Add 0.5 μl 10 mg/ml Proteinase K (Invitrogen), and incubate at 37 °C for 10 min.
5. Transform 2 μl of the reaction mix into 25 μl of either One Shot TOP10 Chemically Competent *E. coli* (Invitrogen) or laboratory-prepared, high-efficiency electrocompetent DH5α

E. coli. Follow standard procedures for transformation, except that incubate for 1.5 h following recovery from heat shock.

6. Spread 150 μl of transformation mix onto LB agar plates containing 75 μg/ml ampicillin.
7. Incubate overnight at 37 °C, and screen three resulting colonies for the correct, fully recombined miniCoopR-GOI clone.

3.3 Making and Selecting Transgenic Animals

Transgenic animals are made by first injecting miniCoopR-GOI clones into *Tg(mitfa:BRAF*^{V600E}*)*; *p53(lf)*; *mitfa(lf)* mutants into single-cell embryos (*see* **Note 5**).

1. Prepare injection mix containing 25 ng/μl miniCoopR-GOI clone, 25 ng/μl capped Tol2 transposase mRNA, and 100 mM KCl in nuclease-free water.
2. Make a microinjection needle using a borosilicate capillary and needle puller. The needle puller heat and time should be set to produce needles that gradually taper to the tip.
3. Load 3 μl of injection mix into the open end of the pulled borosilicate needle.
4. Once the injection mix has settled to the tip of the pulled needle, secure the needle to the injector arm of a micromanipulator.
5. Using a scalpel, break the tip of the needle to yield a beveled edge and a bore size of approximately 5–10 μm.
6. Calibrate the injection volume. Do this by placing a drop of mineral oil on top of a stage micrometer with increments of 0.01 mm or less. Eject liquid into oil, and note the diameter of the oil drop. The desired size is a diameter of 125 μm, which corresponds to a volume of 1 nl. Adjust injection pressure and time settings on the injection apparatus to achieve the proper diameter of 125 μm ejected by a single pulse.
7. Initiate breedings of *Tg(mitfa:BRAF*^{V600E}*)*; *p53(lf)*; *mitfa(lf)* animals, and obtain newly fertilized embryos. During the single-cell stage (0–45 min postfertilization at room temperature), inject into the cell of each embryo with a single pulse of 1 nl of injection mix.
8. After completing injections, place embryos at 28 °C. On the same day as injections are performed, but after embryos are >4 h postfertilization (hpf), remove any embryos that are unfertilized or show severe defects in early developmental cell divisions. Do the same the following day at 24–36 hpf.

3.4 Selection of Melanocyte-Rescued Transgenic Animals

1. At 3 days postfertilization (dpf) place a Petri dish of injected embryos on a white background and view under a dissecting microscope using incident light. Select embryos with melanocyte rescue.

2. At 4 dpf place melanocyte-rescued embryos into the zebrafish facility nursery and raise until 8 weeks of age.
3. At 8 weeks of age, select animals with at least one patch of melanocyte rescue that is greater than 4 mm^2. The melanocytes visible at this time arise between 2 and 4 weeks of age and are termed metamorphic or adult melanoctyes. There is a strong correlation between embryonic melanocyte rescue and rescue of adult melanocytes, most likely because these classes share a common progenitor that is rescued by injected miniCoopR clones. Consequently, selection of embryonic melanocyte-rescued animals yields most if not all animals that will display adult melanocyte rescue.
4. Place 8-week-old rescued animals in tanks at a density of four fishes per liter.

3.5 Screening for Melanoma Onset and Studies of Tumor Tissue

Melanomas arise externally, and their onset is scored by visual inspection of melanocyte-rescued fish. Tumor incidence curves are generated, and Kaplan–Meier analysis is conducted to determine if the gene of interest modifies melanoma onset as compared to control genes. Melanoma-bearing fish can be utilized for a variety of studies, including tumor invasion, genomic profiling, and transplantability.

1. Beginning at 8 weeks of age, monitor melanocyte-rescued fish for the presence of melanomas. The transition from a benign patch of melanocytes to a melanoma involves cell proliferation that raises the lesion so that it protrudes slightly from the surface of the animal. Independent review by trained dermatopathologists (S. Granter and A. Deng, personal communication) indicates that even slightly raised protrusions are malignant whereas melanocyte patches that remain in the same plane as normal skin are benign. Animals with raised protrusions, i.e., melanomas, should be isolated to avoid biting of the lesions by tankmates.

2a. Sectioning and hematoxylin/eosin staining are used to confirm the diagnosis of melanoma and to assay the extent of melanoma cell invasion into the underlying tissue. For the most robust invasion results, only consider fish with melanomas arising on the same anatomical site. The dorsum between the posterior edge of the hindbrain and the anterior aspect of the dorsal fin is best. Nearly half of all melanomas arise in this location, so tumors for study can be readily obtained. To process melanoma-bearing animals for invasion analysis, sacrifice the fish 2 weeks after the initial onset and dissect the tumor-bearing portion by cutting sagittally through the entire animal using a razor blade. Fix the animal for >24 h in 4 % paraformaldehyde solution and then transfer to 500 mM EDTA solution to decalcify. After paraffin embedding the tissue, sagittally section the melanoma, obtaining 5 μm sections, one at every 50 μm interval, until the entire lesion is sectioned. Invasion into dorsal musculature and the spinal column can be quantified using this approach.

2b. To obtain tumor tissue for molecular analyses, sacrifice melanoma-bearing fish at a desired time after tumor onset. Use a razor blade and dissection scissors to obtain tumor tissue and place into extraction media (e.g., Trizol) specific for the type of molecule desired. Use standard procedures for molecular extraction.

2c. For flow cytometry and cell sorting, sacrifice melanoma-bearing fish at a desired time after tumor onset. Use a razor blade and dissection scissors to obtain tumor tissue. Place in 0.9× PBS supplemented with 5 % FBS. Dice the tissue and then triturate extensively to obtain a single-cell suspension. Filter and prepare cell suspension for flow cytometry and cell sorting using established protocols.

3.6 Discussion and Conclusions

The protocols described use attributes of the zebrafish system to facilitate candidate cancer gene screening. The ease of transgenesis and degree of chimerism following injection allow tumor incidence curves to be generated from injected animals rather than making stable transgenic lines for each gene tested. The small size and high fecundity allow each arm of a tumor incidence curve to be built from several animals, enabling studies to be conducted with greater statistical power. These attributes can benefit any cancer genetics study conducted using zebrafish.

There are other aspects of the protocols that are specific to the tissue type being studied. Specifically, the *mitfa(lf)* mutant ablates melanocytes, and the miniCoopR vector is designed to rescue melanocyte development in this mutant. Because the *mitfa* gene acts cell autonomously, each rescued melanocyte contains the *mitfa* minigene and the gene of interest to which it is physically coupled. While these reagents are specific to the pigmentary system, in principle, other tissues could be interrogated by a similar approach. A key reagent is a mutant strain in which the affected gene can be rescued cell autonomously to reconstitute a given tissue. There are several such mutants in zebrafish, affecting both solid tissues as well as a variety of hematopoietic lineages [22, 23], making this strategy viable for additional cell and tumor types.

miniCoopR methodology has been used to identify new genes and mechanisms involved in melanoma progression [24–26]. These studies were conducted by assaying a single gene of interest per injected animal. Given this throughput, roughly 100 genes could be processed over the course of 6 months by a team of four people. However, integration of miniCoopR constructs via Tol2-based transposition results in multiple integration events in a rescued melanocyte. Multiple integration events enable pooled screening, and current injections can be performed with pools of five to eight clones, increasing throughput by a corresponding factor. Using miniCoopR-based screening it is therefore possible to interrogate hundreds of genes for their impact on melanoma progression.

4 Notes

1. Separate entry clones containing a 5′ promoter element, middle open reading frame element, or 3′ element are constructed or obtained from existing plasmid libraries. These clones are then recombined into the miniCoopR vector in a single reaction to generate a miniCoopR-GOI clone.
2. Typically the reverse primer abuts the translational start site of the endogenous gene regulated by the promoter. The forward primer is tailed with sequence containing the *attB4* recombination sequence, and the reverse primer is tailed with sequence containing the *attB1R* recombination sequence. Overall the primers have the following organization:

   Forward primer: 5′-GGGGACAACTTTGTATAGAAAAGTTG + 20–25 nt promoter-specific sequence.

   Reverse primer: 5′-GGGGACTGCTTTTTTGTACAAACTTG + 20–25 nt reverse complement of promoter-specific sequence.
3. The forward and reverse primers are tailed with *attB1* and *attB2* sequences, respectively. In addition, the forward primer is tailed with a Kozak sequence to specify the translational start site. The primers have the following organization:

   Forward primer: 5′-GGGGACAAGTTTGTACAAAAAAGCAGGCTTCGCCACC + 20–25 nt gene-specific sequence, beginning at the initiator methionine codon and extending into the open reading frame.

   Reverse primer: 5′-GGGGACCACTTTGTACAAGAAAGCTGGGTA + 20–25 nt reverse complement of gene-specific sequence, ending at the termination codon (or the penultimate codon if carboxy-terminal tags are going to be fused to the gene of interest) and extending into the open reading frame.
4. The primers for amplification are tailed with *attB2R* and *attB3* sequences and have the following organization:

   Forward primer: 5′-GGGGACAGCTTTCTTGTACAAAGTGG + 20–25 nt 3′ element-specific sequence.

   Reverse primer: 5′-GGGGACAACTTTGTATAATAAAGTTG + 20–25 nt reverse complement of 3′ element-specific sequence.

5. Injection volume must be calibrated precisely so that animals within the same injection group and between different injection groups can be compared. Three days after injection, melanocyte-rescued animals are selected and reared for 8 weeks, at which time a second selection based on melanocyte rescue is performed. An experienced injector should anticipate that approximately 30 % of injected animals will display melanocyte rescue at 8 weeks, so the number of injected animals should be scaled according to the number of melanocyte-rescued animals desired after selection at 8 weeks. In addition, appropriate controls should be performed. miniCoopR-*EGFP* is a standard negative control, whereas miniCoopR-*CCND1* is used as a positive control. *CCND1*, which encodes the CyclinD1 protein, is frequently amplified and overexpressed in human melanomas [20, 21] and can accelerate melanoma onset in the zebrafish (C. Ceol, unpublished results).

Acknowledgments

The authors thank Dr. Yariv Houvras for his contributions in developing miniCoopR-based screening. This work was supported by NIH Pathway to Independence (R00AR056899-03) and American Cancer Society Research Scholar (RSG-12-150-01-DDC) awards to C.C. and a Cancer Research Institute Irvington Fellowship to C.A.P.

References

1. Spitsbergen JM, Tsai HW, Reddy A, Miller T, Arbogast D, Hendricks JD, Bailey GS (2000) Neoplasia in zebrafish (Danio rerio) treated with N-methyl-N′-nitro-N-nitrosoguanidine by three exposure routes at different developmental stages. Toxicol Pathol 28(5):716–725
2. Spitsbergen JM, Tsai HW, Reddy A, Miller T, Arbogast D, Hendricks JD, Bailey GS (2000) Neoplasia in zebrafish (Danio rerio) treated with 7,12-dimethylbenz[a]anthracene by two exposure routes at different developmental stages. Toxicol Pathol 28(5):705–715
3. Langenau DM, Traver D, Ferrando AA, Kutok JL, Aster JC, Kanki JP, Lin S, Prochownik E, Trede NS, Zon LI, Look AT (2003) Myc-induced T cell leukemia in transgenic zebrafish. Science 299(5608):887–890. doi:10.1126/science.1080280, 299/5608/887 [pii]
4. Patton EE, Widlund HR, Kutok JL, Kopani KR, Amatruda JF, Murphey RD, Berghmans S, Mayhall EA, Traver D, Fletcher CD, Aster JC, Granter SR, Look AT, Lee C, Fisher DE, Zon LI (2005) BRAF mutations are sufficient to promote nevi formation and cooperate with p53 in the genesis of melanoma. Curr Biol 15(3):249–254. doi:10.1016/j.cub.2005.01.031, S0960982205000916 [pii]
5. Nguyen AT, Emelyanov A, Koh CH, Spitsbergen JM, Parinov S, Gong Z (2012) An inducible kras(V12) transgenic zebrafish model for liver tumorigenesis and chemical drug screening. Dis Model Mech 5(1):63–72. doi:10.1242/dmm.008367
6. Li Z, Huang X, Zhan H, Zeng Z, Li C, Spitsbergen JM, Meierjohann S, Schartl M, Gong Z (2012) Inducible and repressable oncogene-addicted hepatocellular carcinoma in Tet-on xmrk transgenic zebrafish. J Hepatol 56(2):419–425. doi:10.1016/j.jhep.2011.07.025
7. Haramis AP, Hurlstone A, van der Velden Y, Begthel H, van den Born M, Offerhaus GJ, Clevers HC (2006) Adenomatous polyposis coli-deficient zebrafish are susceptible to

digestive tract neoplasia. EMBO Rep 7(4): 444–449. doi:10.1038/sj.embor.7400638

8. Park SW, Davison JM, Rhee J, Hruban RH, Maitra A, Leach SD (2008) Oncogenic KRAS induces progenitor cell expansion and malignant transformation in zebrafish exocrine pancreas. Gastroenterology 134(7):2080–2090. doi:10.1053/j.gastro.2008.02.084
9. Gill JA, Lowe L, Nguyen J, Liu PP, Blake T, Venkatesh B, Aplan PD (2010) Enforced expression of Simian virus 40 large T-antigen leads to testicular germ cell tumors in zebrafish. Zebrafish 7(4):333–341. doi:10.1089/zeb.2010.0663
10. Neumann JC, Chandler GL, Damoulis VA, Fustino NJ, Lillard K, Looijenga L, Margraf L, Rakheja D, Amatruda JF (2011) Mutation in the type IB bone morphogenetic protein receptor Alk6b impairs germ-cell differentiation and causes germ-cell tumors in zebrafish. Proc Natl Acad Sci U S A 108(32):13153–13158. doi:10.1073/pnas.1102311108
11. Choorapoikayil S, Kuiper RV, de Bruin A, den Hertog J (2012) Haploinsufficiency of the genes encoding the tumor suppressor Pten predisposes zebrafish to hemangiosarcoma. Dis Model Mech 5(2):241–247. doi:10.1242/dmm.008326
12. Gutierrez A, Snyder EL, Marino-Enriquez A, Zhang YX, Sioletic S, Kozakewich E, Grebliunaite R, Ou WB, Sicinska E, Raut CP, Demetri GD, Perez-Atayde AR, Wagner AJ, Fletcher JA, Fletcher CD, Look AT (2011) Aberrant AKT activation drives well-differentiated liposarcoma. Proc Natl Acad Sci U S A 108(39):16386–16391. doi:10.1073/pnas.1106127108
13. Langenau DM, Keefe MD, Storer NY, Guyon JR, Kutok JL, Le X, Goessling W, Neuberg DS, Kunkel LM, Zon LI (2007) Effects of RAS on the genesis of embryonal rhabdomyosarcoma. Genes Dev 21(11):1382–1395. doi:10.1101/gad.1545007, gad.1545007 [pii]
14. Kawakami K, Shima A, Kawakami N (2000) Identification of a functional transposase of the Tol2 element, an Ac-like element from the Japanese medaka fish, and its transposition in the zebrafish germ lineage. Proc Natl Acad Sci U S A 97(21):11403–11408. doi:10.1073/pnas.97.21.11403
15. Peterson RT, Link BA, Dowling JE, Schreiber SL (2000) Small molecule developmental screens reveal the logic and timing of vertebrate development. Proc Natl Acad Sci U S A 97(24):12965–12969. doi:10.1073/pnas.97.24.12965, 97/24/12965 [pii]
16. White RM, Cech J, Ratanasirintrawoot S, Lin CY, Rahl PB, Burke CJ, Langdon E, Tomlinson ML, Mosher J, Kaufman C, Chen F, Long HK, Kramer M, Datta S, Neuberg D, Granter S, Young RA, Morrison S, Wheeler GN, Zon LI (2011) DHODH modulates transcriptional elongation in the neural crest and melanoma. Nature 471(7339):518–522. doi:10.1038/nature09882, nature09882 [pii]
17. Lister JA, Robertson CP, Lepage T, Johnson SL, Raible DW (1999) nacre encodes a zebrafish microphthalmia-related protein that regulates neural-crest-derived pigment cell fate. Development 126(17):3757–3767
18. Kwan KM, Fujimoto E, Grabher C, Mangum BD, Hardy ME, Campbell DS, Parant JM, Yost HJ, Kanki JP, Chien CB (2007) The Tol2kit: a multisite gateway-based construction kit for Tol2 transposon transgenesis constructs. Dev Dyn 236(11):3088–3099. doi:10.1002/dvdy.21343
19. Villefranc JA, Amigo J, Lawson ND (2007) Gateway compatible vectors for analysis of gene function in the zebrafish. Dev Dyn 236(11):3077–3087. doi:10.1002/dvdy.21354
20. Sauter ER, Yeo UC, von Stemm A, Zhu W, Litwin S, Tichansky DS, Pistritto G, Nesbit M, Pinkel D, Herlyn M, Bastian BC (2002) Cyclin D1 is a candidate oncogene in cutaneous melanoma. Cancer Res 62(11):3200–3206
21. Curtin JA, Fridlyand J, Kageshita T, Patel HN, Busam KJ, Kutzner H, Cho KH, Aiba S, Brocker EB, LeBoit PE, Pinkel D, Bastian BC (2005) Distinct sets of genetic alterations in melanoma. N Engl J Med 353(20):2135–2147. doi:10.1056/NEJMoa050092, 353/20/2135 [pii]
22. Haffter P, Granato M, Brand M, Mullins MC, Hammerschmidt M, Kane DA, Odenthal J, van Eeden FJ, Jiang YJ, Heisenberg CP, Kelsh RN, Furutani-Seiki M, Vogelsang E, Beuchle D, Schach U, Fabian C, Nusslein-Volhard C (1996) The identification of genes with unique and essential functions in the development of the zebrafish, Danio rerio. Development 123:1–36
23. de Jong JL, Zon LI (2005) Use of the zebrafish system to study primitive and definitive hematopoiesis. Annu Rev Genet 39:481–501. doi:10.1146/annurev.genet.39.073003.095931
24. Iyengar S, Houvras Y, Ceol CJ (2012) Screening for melanoma modifiers using a zebrafish autochthonous tumor model. J Vis Exp 69:e50086. doi:10.3791/50086
25. Ceol CJ, Houvras Y, Jane-Valbuena J, Bilodeau S, Orlando DA, Battisti V, Fritsch L, Lin WM, Hollmann TJ, Ferre F, Bourque C, Burke CJ, Turner L, Uong A, Johnson LA, Beroukhim R, Mermel CH, Loda M, Ait-Si-Ali S, Garraway LA, Young RA, Zon LI (2011) The histone

methyltransferase SETDB1 is recurrently amplified in melanoma and accelerates its onset. Nature 471(7339):513–517. doi:10.1038/nature09806, nature09806 [pii]

26. Lian CG, Xu Y, Ceol C, Wu F, Larson A, Dresser K, Xu W, Tan L, Hu Y, Zhan Q, Lee CW, Hu D, Lian BQ, Kleffel S, Yang Y, Neiswender J, Khorasani AJ, Fang R, Lezcano C, Duncan LM, Scolyer RA, Thompson JF, Kakavand H, Houvras Y, Zon LI, Mihm MC Jr, Kaiser UB, Schatton T, Woda BA, Murphy GF, Shi YG (2012) Loss of 5-hydroxymethylcytosine is an epigenetic hallmark of melanoma. Cell 150(6):1135–1146. doi:10.1016/j.cell.2012.07.033

Chapter 13

Clonal Screens to Find Modifiers of Partially Penetrant Phenotypes in *C. elegans*

Michael E. Hurwitz

Abstract

Unbiased genetic screens are an excellent way to discover novel genes involved in specific biological processes in vivo. Modifier screens, whether to suppress or enhance a phenotype, are a powerful way to find proteins that modulate biological processes responsible for specific phenotypes. However, modification of phenotypes that are only partially penetrant, which is often the case, are often extremely difficult to screen this way in a traditional F_2 or non-clonal genetic screen. Here we describe an F_3 or clonal screen in the nematode *Caenorhabditis elegans* to search for genes that modify partially penetrant phenotypes. Specifically we describe a screen to search for modifiers of genes that cause defects in migration of a specific developmentally regulated cell, the distal tip cell.

Key words *C. elegans*, Genetic screen, Clonal, Modifier, Distal tip cell

1 Introduction

Genetic screens in invertebrate metazoans have contributed an enormous amount to our understanding of the basic biology of all living things. Some representative examples are the discoveries of the core proteins in programmed cell death [1] and Ras signaling pathways [2, 3].

In a classical genetic screen, animals are exposed to a mutagen, allowed to mate and then either their progeny are analyzed for an abnormal phenotype (in the case of dominant mutations) or the progeny's progeny are analyzed (in the case of recessive mutations). Once animals are isolated that have the phenotype of interest, the mutations are mapped and then causative mutations are identified [4].

Throughout the 1980s and 1990s, many genes were discovered using such screens and then molecularly cloned and analyzed. However, enthusiasm for these types of screens, especially in organisms such as flies and worms waned over time for several reasons. First, many of the easiest and most obvious screens have been done.

Narendra Wajapeyee (ed.), *Cancer Genomics and Proteomics: Methods and Protocols*, Methods in Molecular Biology, vol. 1176, DOI 10.1007/978-1-4939-0992-6_13, © Springer Science+Business Media New York 2014

Second, the rise of RNA interference-based assays in mammalian cells has made large-scale screens in mammalian cells for cell-based phenotypes possible. Third, genetic mapping of mutations can be a laborious, frustrating, and time-consuming process that does not always work.

But the pendulum has swung back to some degree in the worm *Caenorhabditis elegans* because of the development of several techniques that simplify and dramatically speed up identification and validation of mutations obtained in screens. First, high-resolution single nucleotide polymorphism (SNP) maps were used by the Jorgensen group to develop a SNP-based tool to quickly map mutations to specific chromosomes [5]. Second, the Hobert lab has developed methods based on deep sequencing of mutant worm DNA to identify specific mutations [6–8]. The latter method relies on rapid and cheap whole-genome sequencing, which has developed over the last few years. Such sequencing almost certainly will only get cheaper. Third, validation of mutant phenotypes has been simplified by the use of RNA interference, which is fast and easy in *C. elegans* [9, 10].

Standard F_2 screens in *C. elegans* aim to identify recessive mutations in a specific process [4]. Conveniently, *C. elegans* are hermaphroditic so animals contain both sperm and eggs and therefore do not need to be mated to other animals to perform these experiments. In these screens, worms from the parental generation (P_0) are mutagenized and allowed to lay eggs. These eggs (the F_1 generation) are heterozygous for any particular mutation. Thus, to see the abnormal recessive phenotypes, the next generation of animals (F_2) is analyzed. Individual worms that display the phenotype of interest are picked away from the large number of normal worms. F_2 screens allow one to screen a very large number of mutations quickly.

However, F_2 screens have a number of important limitations. They will not discover most lethal mutations because dead animals cannot be isolated. Nor will they discover many mutations that partially modify a phenotype. Lastly, because the nature of screens is to find rare animals ($\leq$1 in 1,000) that look different from the baseline population mutagenized, screens to suppress or enhance a given phenotype (i.e., modifier screens) require the baseline strain to have a fully penetrant phenotype. If even 1 % of animals in the starting strain do not display the baseline phenotype, these animals will swamp out the discovery of any suppressing mutants.

One way to discover mutations that modify a partially penetrant phenotype is to screen for mutants in the F_3 generation. In such a screen, after the P_0 lay eggs, the F_1 animals are allowed to grow up and lay their own eggs. When those eggs hatch and grow up, the F_2 animals are plated individually (i.e., each F_2 gets its own plate) and their progeny (the F_3 generation) are analyzed for the defect. Because all the worms on each plate are potentially homozygous for a recessive mutation that modifies the initial phenotype,

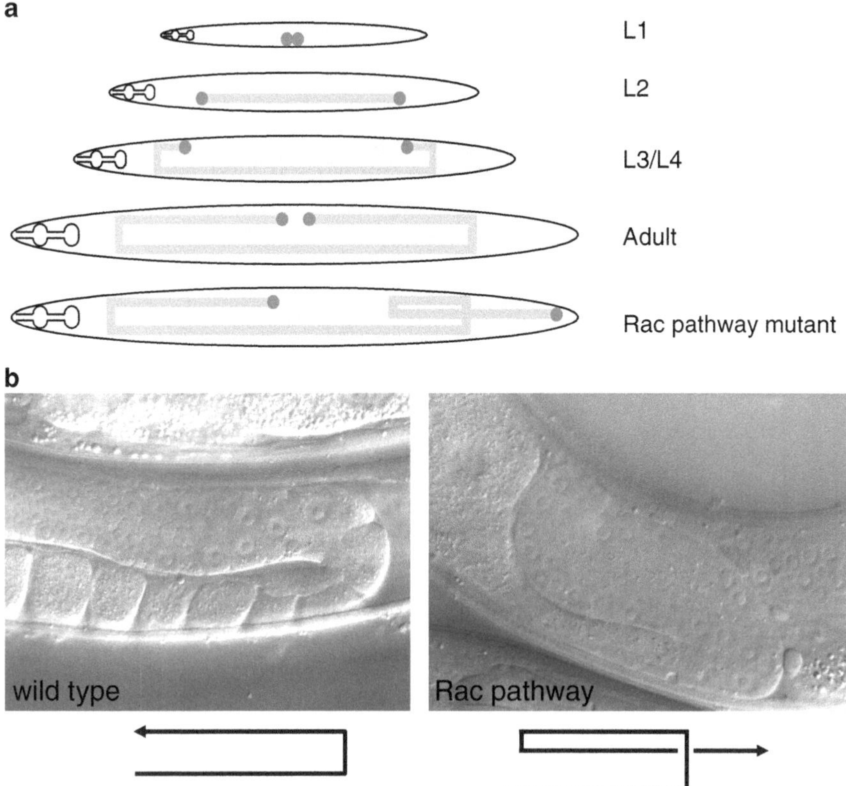

Fig. 1 *C. elegans* gonad development. (**a**) Distal tip cells (DTCs are circles (*red*); gonads are lines (*green*). The two DTCs are in the middle of the animal during the first larval stage (L1) and migrate as shown throughout development. Worms with mutations in genes of the Rac pathway have partially penetrant abnormal DTC migration (both the anterior and posterior gonads can have abnormalities though only the posterior gonad is shown here). The most common phenotype is shown; an extra turn is seen. (**b**) DIC images of wild-type and Rac pathway mutant gonads are shown with the path the DTC took during development shown below schematically. The mutant shown is *ced-2(n1994)* (Color figure online)

one can detect the modification by looking at the whole population. Screens in which a group of animals homozygous for a mutation are analyzed together are called clonal screens.

To illustrate the use of a modifier screen, I will use the following example. *C. elegans* hermaphrodites have two U-shaped gonads. They are formed throughout larval development when the distal tip cells (DTCs), which initially are in the middle of the animal, migrate out to the front and back of the animal, then make a 180° turn and migrate back to the center of the animal [11]. As they migrate, the gonad forms behind them (Fig. 1). Multiple screens have identified a large number of mutations that cause defective DTC migration. One set of genes encodes actin cytoskeletal regulatory proteins [12]. However, the penetrance of the phenotypes is almost never 100 %. In fact, these mutations cause 30–50 % of the gonads to be abnormal.

For example, the gene *ced-5* encodes a component of a heterodimeric guanine nucleotide exchange factor that regulates *C. elegans* Rac, which is required to regulate the actin cytoskeleton and hence required for appropriate DTC migration [13]. *ced-5* mutant animals have approximately 45 % abnormal DTC migration [14]. Since each animal has two gonads, on average only ~30 % of animals will have two normal gonads. To find genes that modify a *ced-5* mutant, we would mutagenize P_0 animals and search in the F_3 generation for a decreased or increased ratio of abnormal DTCs compared to the original strain. The DTC migration pattern can be discerned by examining the shape of the gonads, since the gonads form behind the DTCs as they migrate. However, differential interference contrast (DIC) microscopy is required to visualize the gonad so worms would need to be mounted on a slide to analyze the DTC phenotype this way, which would make the screen exceedingly laborious and inefficient. To circumvent this problem, we have expressed the Green Fluorescent Protein (GFP) under control of the DTC-specific *lag-2* promoter [15]. In wild-type animals, GFP is seen in two spots next to each other but in mutant animals, the two spots are far from each other since the DTCs are not both in the middle of the animal (Figs. 1 and 2).

Below is a detailed description of a clonal F_3 screen for finding modifiers of DTC migration defects. While the details of this screen are optimized for discovery of DTC migration modifiers, aspects of the screen can be adjusted to find other phenotypes.

2 Materials

1. EMS liquid—(Sigma-Aldrich #M0880). For choice of mutagen, *see* **Note 1**.
2. Potassium hydroxide pellets.
3. Glass pipets with rubber bulbs.
4. OP50 *E. coli* bacteria.
5. Dissecting fluorescence microscope.
6. Worm pick.
7. M9 medium: Put the following ingredients in a 2 l flask: 5.8 g Na_2HPO_4, 3.0 g KH_2PO_4, 0.5 g NaCl, and 1.0 g NH_4Cl. Bring to 1 l with distilled water and stir until dissolved. Dispense 100 ml into bottles (ten bottles) and autoclave with lids on but not tightened. Once lids are tightened (after autoclaving), M9 can be stored indefinitely.
8. Potassium phosphate, pH 6.0: Make the following two solutions separately. (a) 1 M KH_2PO_4: dissolve 102.1 g KH_2PO_4 in water and adjust volume to 750 ml. (b) 1 M K_2HPO_4: dissolve 52.25 g K_2HPO_4 in water and adjust volume to 300 ml. Then add K_2HPO_4 solution to KH_2PO_4 solution until the pH is brought

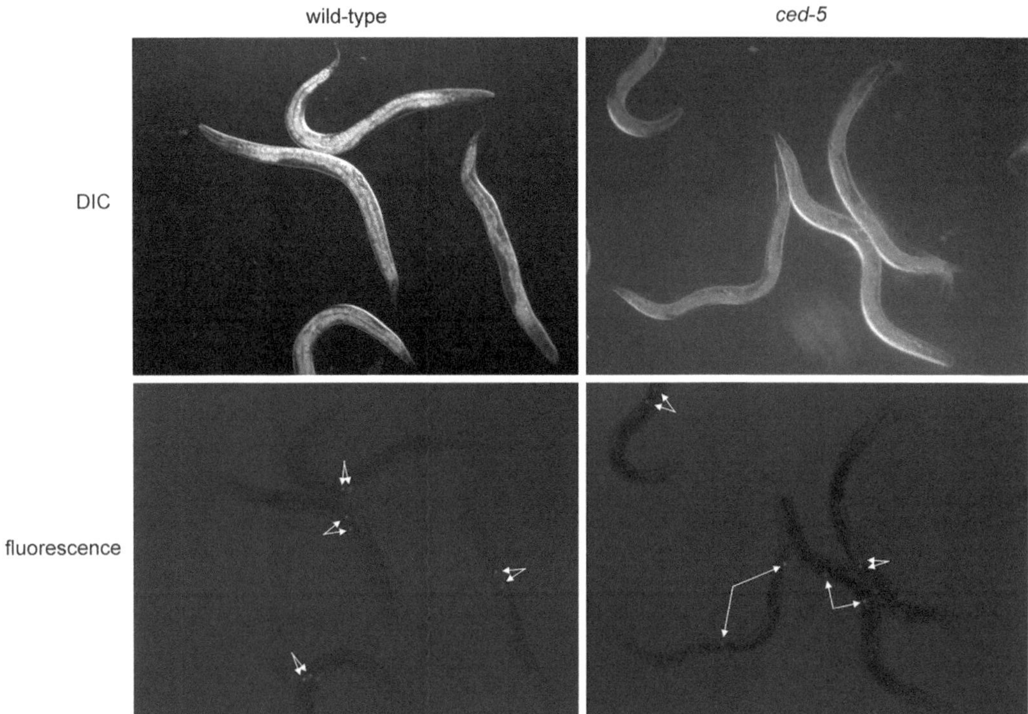

Fig. 2 *lag-2::gfp* in wild-type and Rac pathway mutant animals. The Rac pathway mutant has the genotype *ced-5(n1812)*. *lag-2::gfp* is expressed in the DTCs, which are indicated by *arrows*. The space between the DTCs is increased in mutant animals. Note that the phenotype is partially penetrant (only two of the four animals have abnormally located DTCs)

from ~4.5 to 6.0 (takes ~150 ml of K_2HPO_4). Then adjust volume to 1 l. Autoclave with lid on but not tightened. Potassium phosphate can be stored indefinitely.

9. NGM plates: Put the following ingredients into a 4 l Erlenmeyer flask: 975 ml distilled water, 3.0 g NaCl, 2.5 g peptone, 1 ml cholesterol (5 mg/ml in ethanol). Mix until dissolved by vigorous shaking. Add 17.0 g agar. Autoclave. When the temperature is below 70 °C, add the following: 1 ml 1 M $CaCl_2$, 1 ml 1 M $MgSO_4$, and 25 ml potassium phosphate pH 6.0. Pour plates. Each 6 cm plate requires approximately 8 ml agar. Each 10 cm plate requires approximately 25 ml agar. Plates can be stored at 4 °C for weeks. If $CaCl_2$ is added when the temperature is above 70 °C, the calcium will come out of solution and the agar will be much less translucent.
10. B broth: Dissolve 5 g Bacto Tryptone and 2.5 g NaCl in 450 l water and pH to 7.0 with NaOH. Adjust volume to 500 ml. Dispense 100 ml into bottles and autoclave with lids on but not tightened. Once lids are tightened (after autoclaving), B broth can be stored indefinitely.

The basic methods of *C. elegans* culture were described in the landmark paper by Sydney Brenner in 1974 [16] and have not changed substantially since then. Worms are grown on agar plates with an auxotroph *E. coli* strain (OP50 or HB101) as the food source (so that the *E. coli* does not grow too thick, which would obscure observing the worms on the plate). To move worms, they can be washed off of plates en masse with M9 or they can be individually picked with a "worm pick," a ~2 cm long piece of platinum wire, crimped at the end with pliers and mounted on a handle of some sort (we usually use a Pasteur pipet with the end melted onto the metal). The platinum pick is placed in the flame of a Bunsen burner to sterilize it but because it cools so rapidly, it can be used almost immediately after being exposed to flame to move animals.

To perform these types of screens, the operator will also need to know basic techniques of worm handling, including an understanding of the life cycle, its timing and how to manipulate that. Much of this information can be found in the aforementioned paper by Brenner [16], in two textbooks, *C. elegans* I [17] and *C. elegans* II [18] and online at www.wormbook.org.

3 Methods

3.1 Mutagenesis

1. Grow up mixed-stage worms on several 6 cm NGM plates (*see* **Note 2** for seeding of bacteria on plates), making sure that a lot of mid-late L4 (fourth larval stage) worms are on the plates.
2. Wash animals off the plate with 5 ml M9, and using a glass pipet, transfer them to a 15 ml centrifuge tube. Worms adhere to plastic pipets so glass pipets are necessary.
3. Pellet worms in a tabletop centrifuge at 1,000 rpm for 30 s and remove enough supernatant so that the total volume is 3 ml.
4. In a 15 ml tube add 20 μl of EMS to 1 ml of M9 and shake to resuspend. All steps that require handling of EMS should be performed in a hood (*see* **Note 3** for EMS handling).
5. Transfer 3 ml of worms in M9 to the 15 ml tube containing the EMS suspension. The final concentration of EMS in the 4 ml is 50 mM.
6. After closing tube tightly, double wrap lid with Parafilm. Then gently rock at 20 °C for 4 h.
7. After incubation pellet worms as before (1,000 rpm for 30 s). Remove supernatant to another tube taking care not to disturb the worm pellet, which is not tightly packed. Then wash twice with 3 ml M9 (pelleting worms in between) and remove supernatant to the same tube used for the first supernatant.

Add five to six pellets of potassium hydroxide (KOH) to the EMS-containing supernatants to neutralize it. KOH treatment detoxifies EMS (*see* **Note 3**).

8. Using a glass pipet, transfer the worms to an NGM plate containing bacteria and let the worms recover for 1–2 h.

3.2 Screen

1. Pick 80–100 young adults to several 6 cm NGM plates and leave them at 20 °C overnight. The eggs laid overnight are not used. Some of these eggs might have been in meiosis during the mutagenesis, potentially resulting in loss of the mutations by DNA repair. Each plate should be numbered and which plate worms of subsequent generations come from should be noted. That way, if, for example, ten mutant strains are isolated and eight of them came from the same P_0, it is likely that those eight are not independent mutants. Rather they probably arise from a "jackpot" (*see* **Note 4**).
2. Transfer four gravid adult worms to 20–30 10 cm NGM plates and let them lay approximately 50 eggs (per plate, not per animal). This should take approximately 3–4 h. *See* **Note 5** for numbers of worms and plates to use.
3. Remove the adults (P_0) from the plates by picking and discard them. These eggs are the F_1 generation. Note: the number of eggs laid depends on the specific genotype of the animals in the screen. The time required for laying the desired number of eggs needs to be tested for each genotype before doing the screen. In addition, often after mutagenesis the animals are sicker and lay fewer eggs.
4. Let the F_1 animals grow up and lay eggs. Once the F_1 animals have laid eggs for several hours, remove the F_1 adults from the plate. The simplest way to do this is to wash the plate with M9 and suck the M9 away with a suction apparatus connected to a glass pipet with a 100 μl pipet on its tip. The eggs will stick to the plate while the adult F_1 animals are washed away.
5. Take one half of the plates and incubate at 20 °C. Incubate the other half of the plates at 16 °C. The animals incubated at 20 °C will reach adulthood approximately 45 h after hatching; the animals incubated at 16 °C will reach adulthood approximately 72 h later (*see* Table 1 and **Note 6**). This will allow the person doing the screening to do so over 2 days, which will probably be required to complete the screen.
6. When the next generation of animals (the F_2 generation) is at the late L4 stage or early adulthood, transfer 25 animals from each plate to their own 6 cm plates (i.e., one F_2 animal per plate). This will require approximately 500 plates for a small pilot screen in which approximately 400 independent haploid genomes are analyzed (*see* **Notes 5**, **7** and **8**). At this point,

Table 1
Temperature dependence of development in wildtype *C. elegans*

T (°C)	L1 (h)	L2 (h)	L3 (h)	L4 (h)
16	23	39	53	72
20	14	24	33	45
25	11	19	25	34

The table indicates the number of hours after hatching that each stage ends at the temperature indicated

incubate all plates at 23 °C. *See* **Note 8** for why 25 F_2 animals are picked from a plate of 50 F_1 animals.

7. After 24 h of laying, remove and discard the adults.
8. When the progeny (F_3) reach adulthood in approximately 48 h, perform screen on the dissecting fluorescence microscope.
9. Plates that have animals with either significantly more or fewer abnormal gonads should be saved and progeny from those plates rechecked with formal quantification of gonad defects to validate that the difference was not caused by chance.

Once mutants are isolated, gene mapping and direct sequencing can be performed by the methods mentioned in the introduction.

4 Notes

1. Choice of mutagens: there are a number of different ways to mutagenize *C. elegans*, including chemical mutagens (EMS and ENU), UV radiation or transposon-based approaches. The most commonly used and easiest is chemical mutagenesis, usually done with methanesulfonic acid, ethyl ester (EMS), which primarily causes transitions (G/C-A/T) and occasionally deletions, especially at higher concentrations [19, 18].
2. Bacterial seeding of NGM plates: NGM plates are seeded with bacteria as the food source for the worms. OP50 bacteria on a plate are re-streaked onto a new NGM plate and a single colony is picked with a sterile loop and placed into 100 ml B broth in the bottle. Because no antibiotics are used, care should be taken to do this in as sterile a fashion as possible. B broth is incubated at 37 °C overnight (no need for shaking) and then can be stored in a refrigerator (4 °C) for up to 6 weeks. To seed plates, fill a 10 ml pipet with OP50 in B broth and drop one to two drops onto each 6 cm plate or three to four drops onto each 10 cm plate. Plates can be shaked slightly to increase the

size of the drop on the plate depending on the desired size of the bacterial lawn. Incubate plates overnight at 37 °C and then let cool to room temperature (20–23 °C) before use. These plates can be stored at room temperature in a covered box for up to 2 weeks.

3. EMS disposal: EMS is extremely toxic and should only be handled in a fume hood. We double glove and have strict procedures for disposal, including special containers for all gloves, pipet tips and other supplies that touch the EMS. We use a dedicated micropipettor also. The only supplies we do not dispose of this way are tubes containing EMS that has been neutralized with potassium hydroxide as described above. Tubes with EMS-containing liquids are also decontaminated by the neutralization process with potassium hydroxide.
4. Stage of animal development in which to mutagenize: when mutagenizing animals, it is important to choose animals at the right stage. Late fourth larval stage (L4) animals or young adults are the optimal age for mutagenesis. Younger animals have comparatively few germline nuclei. Mutagenesis of these younger animals theoretically can result in mutation of a small number of nuclei, which will then divide multiple times and become gametes, resulting in many progeny with the same mutations (referred to as "jackpots"), rather than progeny harboring independent mutations. Older adults are not as hardy and will be killed by the mutagenesis or have a very decreased brood size.
5. Number of animals and plates required: for most F_3 screens, because we need to analyze whole populations, smaller numbers of genomes are analyzed, usually only 500–1,000 genomes. Although there is no limit to the number one can analyze, such screens are comparatively time- and labor-intensive. Thus smaller numbers are analyzed. To analyze 500 F_3 worm genomes, we will need to plate 500 F_2 worms individually to plates. For statistical reasons (see below), we will pick 25 F_2 worms from plates that originally contained approximately 50 F_1 worms, so approximately 20 plates containing 50 F1 worms are needed. If 4 P_0 animals are put on each plate, only 80 P_0 animals are needed in total.
6. Temperature: genetic screens can be performed at different temperatures to look for temperature-sensitive (either heat or cold) mutants. In addition, the time it takes for worms to proceed through development is highly temperature-dependent (*see* Table 1). For this screen, we incubate some worms at 20 °C and others at 16 °C during the screen so that we do not need to analyze all the animals at once. Once the F_3 generation is laid, we will keep the animals at room temperature (~23 °C), primarily for convenience.

7. Number of genomes analyzed: geneticists usually consider the "number of genomes" analyzed in a genetic screen. For example, each F_1 animal contains two haploid genomes, each with independent mutations. If we analyze F_2 animals for recessive mutations, those will only be seen if they are homozygous. Therefore, in F_2 screens, each animal actually screened only represents one haploid genome. The same is true for F_3 recessive screens; each F_3 represents one haploid genome.
8. Statistical considerations: When picking F_2 progeny from a plate, the goal is to choose animals with independent mutations. Thus, ideally F_2 animals from different F_1 mothers should be chosen. Although it is impossible to know whether F_2 progeny come from the same or different mothers, we can calculate the average number of independent F_2 progeny picked by calculating the probability of picking any single F_1 progeny zero times and subtracting that from one, then multiplying the entire result by the number of F_1 animals. The probability of picking a specific F_1 progeny zero times can be calculated using a binomial coefficient. The calculation is below.

   X = number of F_1 animals

   N = number of F_2 animals picked

   T = number of times a particular animal is picked

   The number of independent F_2 animals picked is:

   $$X(N!/T!(N-T)!) - X(1/X)^{T}(1-1/X)^{N-T}$$

   For $T=0$:

   $$X(N!/0!N!) - X(1/X)^{0}(1-1/X)^{N} = X - X(1-1/X)^{N}$$

   For $X=50$ and $N=25$:

   $50 - 50(0.98)^{25} = 19.8$

   So if we pick 25 F_2 animals from the progeny of 50 F_1 animals on a plate, on average 20 of them will be from independent F_1 animals. As can be seen from the equation, the more animals picked, the more likely that the animals picked will be from the same F_1 parents. For an excellent treatment of statistical issues encountered in genetic screens, see the paper by Shaham [20].

Acknowledgments

I would like to thank Marc Hammarlund and Valerie Reinke for helpful discussions.

References

1. Metzstein MM, Stanfield GM, Horvitz HR (1998) Genetics of programmed cell death in *C. elegans*: past, present and future. Trends Genet 14:410–416
2. Moghal N, Sternberg PW (2003) The epidermal growth factor system in *Caenorhabditis elegans*. Exp Cell Res 284:150–159
3. Rubin GM, Chang HC, Karim F et al (1997) Signal transduction downstream from RAS in *Drosophila*. Cold Spring Harb Symp Quant Biol 62:347–352
4. Jorgensen EM, Mango SE (2002) The art and design of genetic screens: *Caenorhabditis elegans*. Nat Rev Genet 3:356–369
5. Wayne Davis M, Hammarlund M, Harrach T et al (2005) Rapid single nucleotide polymorphism mapping in *C. elegans*. BMC Genomics 6:118
6. Sarin S, Prabhu S, O'Meara MM et al (2008) *Caenorhabditis elegans* mutant allele identification by whole-genome sequencing. Nat Methods 5:865–867
7. Bigelow H, Doitsidou M, Sarin S et al (2009) MAQGene: software to facilitate *C. elegans* mutant genome sequence analysis. Nat Methods 6:549
8. Doitsidou M, Poole RJ, Sarin S et al (2010) *C. elegans* mutant identification with a one-step whole-genome-sequencing and SNP mapping strategy. PLoS One 5:e15435
9. Fire A, Xu S, Montgomery MK et al (1998) Potent and specific genetic interference by double-stranded RNA in *Caenorhabditis elegans*. Nature 391:806–811
10. Timmons L, Fire A (1998) Specific interference by ingested dsRNA. Nature 395:854
11. Kimble J, Hirsh D (1979) The postembryonic cell lineages of the hermaphrodite and male gonads in *Caenorhabditis elegans*. Dev Biol 70:396–417
12. Lehmann R (2001) Cell migration in invertebrates: clues from border and distal tip cells. Curr Opin Genet Dev 11:457–463
13. Wu YC, Horvitz HR (1998) *C. elegans* phagocytosis and cell-migration protein CED-5 is similar to human DOCK180. Nature 392: 501–504
14. Hurwitz ME, Vanderzalm PJ, Bloom L, Goldman J, Garriga G, Horvitz HR, Green DR (2009) Correction: Abl kinase inhibits the engulfment of apoptotic cells in *Caenorhabditis elegans*. PLoS Biol 7(4):e99
15. Henderson ST, Gao D, Lambie EJ et al (1994) *lag-2* may encode a signaling ligand for the GLP-1 and LIN-12 receptors of *C. elegans*. Development 120:2913–2924
16. Brenner S (1974) The genetics of *Caenorhabditis elegans*. Genetics 77:71–94
17. Wood W, the Community of C. elegans Researchers (1988) The nematode *Caenorhabditis elegans*. Cold Spring Harbor Laboratory Press, Cold Spring Harbor, NY
18. Johnsen RC, Baillie DL (1997) Mutation. In: Riddle DL et al (eds) *C. elegans* II. Cold Spring Harbor Laboratory Press, Cold Spring Harbor, NY, pp 79–95
19. Anderson P (1995) Mutagenesis. Methods Cell Biol 48:31–58
20. Shaham S (2007) Counting mutagenized genomes and optimizing genetic screens in *Caenorhabditis elegans*. PLoS One 2:e1117

Chapter 14

Serum Profiling Using Protein Microarrays to Identify Disease Related Antigens

Donald Sharon and Michael Snyder

Abstract

Disease related antigens are of great importance in the clinic. They are used as markers to screen patients for various forms of cancer, to monitor response to therapy, or to serve as therapeutic targets (Chapman et al., Ann Oncol 18(5):868–873, 2007; Soussi et al., Cancer Res 60:1777–1788, 2000; Anderson and LaBaer, J Proteome Res 4:1123–1133, 2005; Levenson, Biochim Biophy Acta 1770:847–856, 2007). In cancer endogenous levels of protein expression may be disrupted or proteins may be expressed in an aberrant fashion resulting in an immune response that bypasses self tolerance (Soussi et al., Cancer Res 60:1777–1788, 2000; Disis et al., J Clin Oncol 15(11):3363–3367, 1997; Molina et al., Breast Cancer Res Treat 51:109–119, 1998). Protein microarrays, which represent a large fraction of the human proteome, have been used to identify antigens in multiple diseases including cancer (Anderson and LaBaer, J Proteome Res 4:1123–1133, 2005; Disis et al., J Clin Oncol 15(11):3363–3367, 1997; Hudson et al., Proc Natl Acad Sci U S A 104(44):17494–17499, 2007; Beyer et al., J Neuroimmunol 242:26–32, 2012). Typically, arrays are probed with immunoglobulin (Ig) samples from patients as well as healthy controls, then compared to determine which antigens (Ag's) are more reactive within the patient group (Hudson et al., Proc Natl Acad Sci U S A 104(44):17494–17499).

Key words Protein microarray, Detection antibody, Immunoglobulin, Cancer antigen, Serum, Fluorescent dye, Nitrocellulose, Quantile normalization

1 Introduction

Protein arrays have become a standard method for assaying interactions of biomolecules on a broad scale. Originally developed to screen protein–protein interactions and enzyme activity [1], these arrays have become standard tools for detecting disease specific antibody (Ab)–antigen interactions in various disease states [2, 3, 4, 5]. Other screening methods like SEREX (serological analysis of cDNA expression libraries) or 2-D PAGE require additional steps to identify the antigens and suffer from poor reproducibility [6]. This is not the case with protein microarrays as each spot is individually addressable and hundreds of arrays can be produced in the same lot resulting in high degree of uniformity.

Narendra Wajapeyee (ed.), *Cancer Genomics and Proteomics: Methods and Protocols*, Methods in Molecular Biology, vol. 1176, DOI 10.1007/978-1-4939-0992-6_14, © Springer Science+Business Media New York 2014

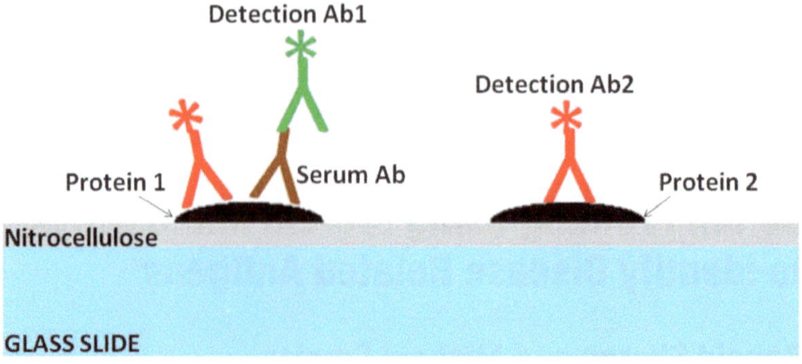

Fig. 1 Cross-sectional diagram of a typical protein array that has been probed with an Ig sample and fluorescently labeled secondary antibodies to the Ig and common epitopes of the arrayed proteins. The glass surface of the array is shown in *blue* and the nitrocellulose surface coating in *gray*. Proteins are represented in *black*. Bound Ig's are depicted in *brown*, while the anti-Ig detection antibodies are *green* and the anti-epitope detection antibodies in *red*. There is a serum response to the protein on the *left*, while the protein to the *right* is not reactive with Ig's in the sample. Both proteins are recognized by the anti-epitope detection antibody as both contain the same fusion tag (e.g., GST)

Most of these arrays are made by printing individually purified proteins or lysates onto glass slides that have been coated with nitrocellulose (FAST and PATH arrays) (Fig. 1). Other surface chemistries are used and different spotting methods such as NAPPA (nucleic acid programmable protein arrays) have been demonstrated as well [7]. The resolution of these arrays is only limited by the feature size (~100 μM diameter for spots on printed arrays) and the number of spots that can be printed on them.

Currently, some commercial arrays feature more than 9,000 full length human proteins (approximately one-third of the proteome) and arrays that represent the entire proteome are expected in the near future. Traditionally there have been three categories of protein microarray: (1) functional arrays, (2) lysate arrays, and (3) capture arrays. Functional protein arrays are printed using full-length proteins or large domains of proteins that retain their native functional activity [1]. Lysate arrays contain gross cellular lysates rather than individually expressed and purified proteins. The last category of arrays, capture or antibody arrays, are printed with antibodies that are meant to bind and pull down targeted proteins in a sample that can then be detected using a specific secondary antibody much like an enzyme linked immunosorbent assay (ELISA) [8]. Proteins that are used in preparation of functional arrays are often produced using a baculovirus system in insect cells (sf9 cells) and include an epitope tag, such as GST, to facilitate purification [2]. One drawback to using proteins expressed in such a system for evaluating human self antigen responses is that they may not display normal levels of posttranslational modifications. Thus, a serum response that is directed against modified residues may be overlooked [9, 2].

Recently, photochemical printing of peptides has caught up to that of their nucleotide microarray counterparts as high throughput sequencing has replaced such arrays and whole proteome arrays can now be produced. Roche is currently developing such an array that will tile the entire proteome at a resolution using overlapping peptides. The resolution is important for epitope mapping (complimentary determining regions of antibodies bind to epitopes of approximately six amino acids). While peptide arrays offer the highest density, there is the disadvantage that some binding interactions that require native protein conformation may not be observed [9, 2, 1].

Cancer antigens have demonstrated incredible importance in the clinic for screening and as prognostic indicators [10, 11, 9, 12]. The experimental design for identifying novel antigens using protein arrays often involves probing the arrays with sera or other body fluids (e.g., synovial fluid, cerebrospinal fluid, or sputum) that contain Ig's that may recognize epitopes of antigens spotted or printed on the arrays. Typically, the arrays are incubated with an appropriate dilution of the Ig (empirically determined) from a diseased group as well as from an age/sex-matched healthy donor group. The arrays are then washed to remove unbound Ig and incubated again with a secondary detection antibody that has been conjugated to a fluorescent dye. Image data is then collected by scanning the arrays using a fluorescent scanner (Fig. 2). Differences in antigen reactivity between the patient and control groups can then be determined. However, this oversimplifies the roles that data normalization and analysis play in these studies. Protein microarrays suffer from greater artifacts and increased background compared to nucleic acid microarrays and since many cancer antigens may show only subtle differences in reactivity it is necessary to normalize to account for artifacts that may obscure those antigens [2]. Effective statistical methods have been developed to smooth away regional artifacts and quantile normalization often is conducted prior to statistical comparison of the two groups [2, 13]. It is often desirable to screen a smaller cohort of patients using more costly high density arrays in order to identify candidate self antigens. These markers may then be validated by screening many more patients for reactivity using more focused assays, as this approach can be more cost-effective and higher throughput. Histological methods such as immunohistochemical (IHC) staining are often employed to validate candidate cancer antigens identified in protein array screens [2].

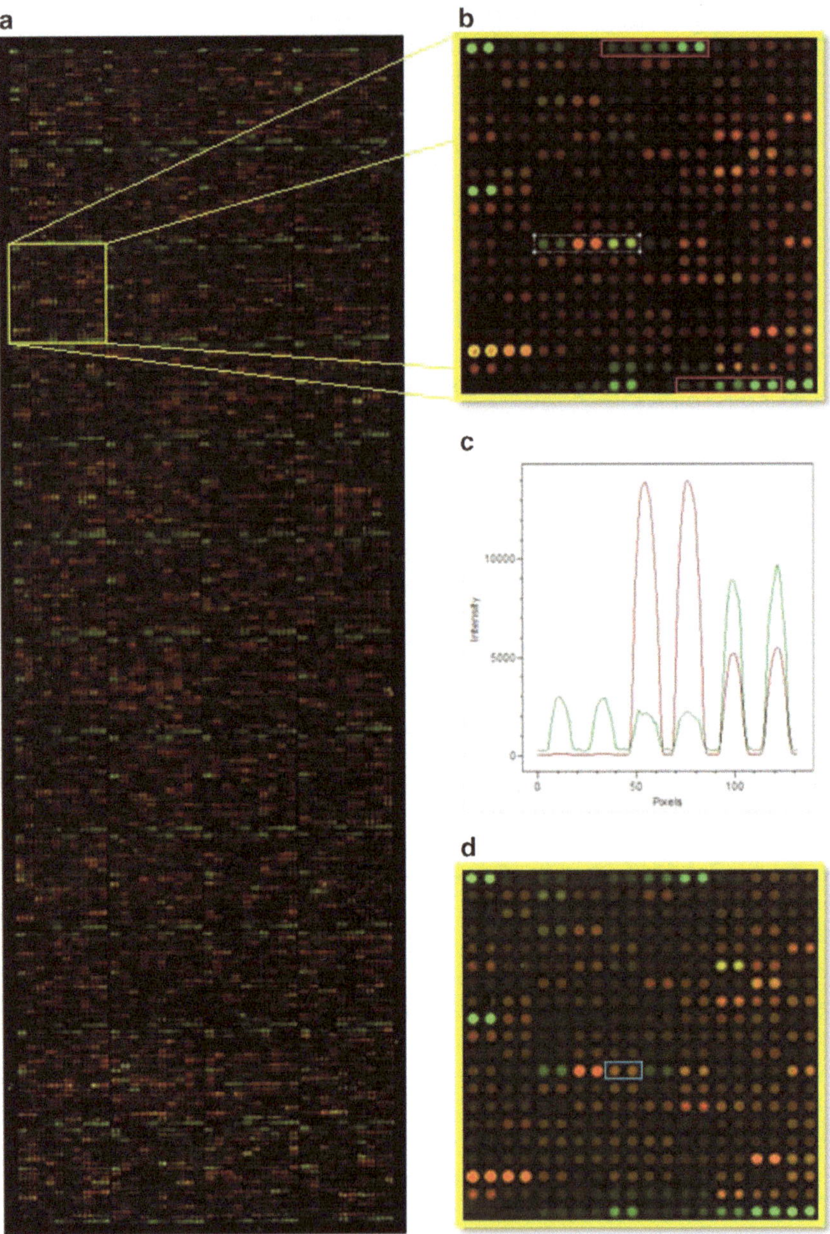

Fig. 2 Panel **a** shows a protein microarray with 9,000+ full length GST tagged human proteins spotted in duplicate that was probed with sera from a patient with multiple myeloma. Human IgG signals were detected with an Alexa Fluor 555 anti-human IgG secondary antibody and GST-protein content was detected with an Alexa Fluor 647 anti-GST antibody. IgG signal is shown in *green* and GST in *red*. The *yellow highlighted* block is shown in panel **b**. IgG and GST reactive protein antigens are visible and dilution series of human IgG and anti-human IgG capture antibodies are shown in *red blocks*. Reactivity profiles of the antigens in the *white block* are shown in panel **c**. Spot reactivity is shown in order from *left* to *right*. Panel **d** shows the same block from a corresponding array that was probed with sera from a healthy individual. An antigen that shows differential reactivity from the myeloma serum probing is shown in the *blue block*

2 Materials

All buffers are prepared using molecular biology grade H_2O. Arrays are typically probed in batches of 20 per day over a period of 4–6 days (80–120 arrays are usually printed in a lot). A 10 % excess of each buffer is prepared and cooled to 4 °C at the start of the procedure each day. If probing a different number of arrays, the buffer volumes that are prepared may be scaled accordingly.

1. Blocking buffer: using a graduated cylinder prepare 110 mL to a final concentration of 50 mM HEPES, 200 mM NaCl, 0.08 % Triton X-100, 25 % glycerol, 20 mM glutathione (reduced), and 1× synthetic block (Life Technologies Inc. catalogue # PA017) (*see* **Note 1**). Then adjust the pH to 7.5 using a 4 M NaOH solution and add water to a final volume of 110 mL (*see* **Note 2**). Finally, add DTT to 1 mM just prior to blocking the arrays.
2. Wash buffer: prepare by making a 1× PBS (pH = 7.4), 1× synthetic block, and 0.1 % Tween-20 solution in a final volume of 1,500 mL using a graduated cylinder (*see* **Note 3**).
3. Patient samples: store human plasma/serum at −80 °C and avoid multiple freeze–thaw cycles. Prepare dilutions on ice with wash buffer within 2 h of probing the arrays.
4. Secondary detection reagents: Alexa Fluor-555 coupled highly anti-human IgG may be purchased commercially. Alexa Fluor 647 coupled anti-GST is also commercially available (*see* **Note 4**).

3 Methods

Always be cautious when working with human plasma or sera, take appropriate precautions and wear personal protective equipment (e.g., lab coat, latex gloves, and goggles). Be sure to become trained in laboratory safety involving bloodborne pathogens as well as general personal safety protocol. This protocol is adapted from our original Protometrix protocols and Invitrogen's Protoarray IRBP protocol and is appropriate for immune response profiling using all FAST/PATH based protein microarrays (i.e., nitrocellulose coated glass slide based arrays).

1. Thaw arrays: Remove arrays from freezer and place in a 4 °C cold room (remainder of protocol up to the water rinse step will be carried out at 4 °C). Allow the arrays to equilibrate for up to 30 min (*see* **Note 5**). Then remove the arrays using forceps or tweezers. Place the arrays into a four-well slide culture tray (*see* **Note 6**).

2. Incubate the arrays with blocking buffer: Add 5 mL of blocking buffer to each well. Gently tilt the trays back and forth until the blocking buffer covers the entire surface of each array. Incubate with gentle shaking (<100 rpm) on a circular shaker for 1 h (*see* **Note 7**). Remove the blocking buffer completely. Quickly proceed to the next step to avoid drying of the array surface (*see* **Note 8**).
3. Rinse the arrays with wash buffer: Add 5 mL of washing buffer to each well. Incubate with shaking as above. After 5 min remove the buffer completely and proceed immediately to the next step (*see* **Note 9**).
4. Serum incubation: Thaw all serum samples on ice. Clarify the serum samples by spinning in a microcentrifuge at 20,000 × *g* (or maximum speed) for 10 min at 4 °C. Dilute serum into 5 mL of wash buffer. Dispense 5 mL of diluted serum into each well (one negative control slide may be incubated with 5 mL of wash buffer without serum). Incubate for 1.5 h with shaking as before (*see* **Note 10**).
5. Wash the arrays: Aspirate the samples from each well. Add 5 mL of wash buffer to the wells and incubate for 5 min with shaking. Repeat the wash four more times (*see* **Note 11**).
6. Incubate the arrays with secondary detection antibodies: Dilute fluorescently labeled anti-human Ig secondary detection antibody in wash buffer. Add 5 mL of the diluted antibody to each well. Incubate arrays with the detection antibody for 1.5 h with rocking at 4 °C in the dark (*see* **Note 12**).
7. Wash the arrays: Aspirate the samples from each well. Add 5 mL of wash buffer to the wells and incubate for 5 min with shaking. Repeat the wash step four more times (*see* **Note 13**). Do not aspirate the buffer after the final wash.
8. Water rinse: Prepare a beaker with a sufficient volume of molecular biology grade H_2O to fully submerge the full length of the arrays. Gently submerge the arrays in the water bath (*see* **Note 14**). This step should be performed at room temperature. Proceed immediately to drying of the arrays.
9. Drying of the arrays: Dry the arrays by placing them into a light block slide box and spinning them at 200 × *g* for 2 min in a swinging bucket type centrifuge with plate adaptor at room temperature (*see* **Note 15**). Allow the slides to stay at room temperature in the slide boxes overnight before scanning (*see* **Note 16**).
10. Data collection: Scanning of the arrays should be performed within 24 h of probing. Scanner should be properly calibrated prior to scanning. All scanner settings should remain the same while scanning the arrays (*see* **Note 17**).
11. Data processing and analysis may be conducted using standard techniques (*see* **Note 18**).

4 Notes

1. Life Technologies offers a premade blocking buffer that can also be used when working with PATH based arrays. Bovine serum albumin (BSA) or Hammerstein casein, at a final concentration of 10 mg/mL, may be substituted for the synthetic blocking agent. Filter the solution using a 500 mL, 0.2 μm vacuum filter unit. This helps to reduce debris that can bind to the arrays increasing background. Take care to avoid foaming of the buffer during filtration (adding the Triton X-100 after filtering can help). It is convenient to prepare a 10 % solution of triton X-100 for use in making the buffer. If the arrays are to be printed in house, or if it is desirable to include additional antigens, purified or commercially purchased antigens may be mixed 50/50 volume/volume with blocking buffer (minus reduced glutathione, blocking agent, and DTT) and spotted onto the array's surface using a manual or automated slide printer.
2. The need to adjust the pH manually may be avoided by using commercial stock solutions of HEPES that has already been pH adjusted. If adjusting pH manually, care should be taken when near the desired pH not to surpass it by switching to a 1 M NaOH solution.
3. A 10× PBS stock solution that is already adjusted to pH = 7.4 may be purchased from many commercial vendors. A 10 % Tween-20 solution should be prepared in advance. Again, other blocking agents (i.e., BSA or casein) may be substituted for the synthetic block. A synthetic blocking reagent has the advantage of a known composition resulting in improved reproducibility. This buffer may be filtered using a 0.2 μm vacuum filter unit.
4. Secondary antibodies should be polyclonal and Alexa Fluor dyes are preferable to Cy dyes as the latter are more sensitive to photodamage.
5. While transporting arrays, keep the containers they are in on ice. Arrays should be kept in sealed slide boxes or slide mailers to avoid exposure to air or moisture until they reach ambient temperature in the cold room. It is important not to remove the arrays from their containers until they are the same temperature as the surrounding air to avoid precipitation from forming on the array surface.
6. Avoid touching the protein containing regions of the array. It is best to handle the arrays from the barcoded end (or the end furthest from the proteins). It is also critical that the arrays be oriented such that the side containing the spotted proteins is facing up in the trays.

7. This step serves to block nonspecific interactions and reduces background (similar to blocking a membrane with blotto in western blotting). Blocking buffer should be dispensed against the sides or corners of the wells, not directly onto the surface of the arrays as this may disrupt or damage the protein spots. The trays may be stacked on top of each other during rocking steps to assure that all of the arrays are shaken at the same speed and orientation.
8. It is preferable to vacuum aspirate the buffer using a pipette tip (it may be necessary to clip the tip or use a tip with a wide bore as the blocking buffer contains glycerol and is highly viscous). To effectively remove the buffer it helps to slightly tilt the trays.
9. This step serves to remove any remaining blocking buffer. If the amount of sample available will not be enough for performing the incubations in 5 mL after dilution in wash buffer, then the slides may be removed from the trays and incubated using other means that are compatible with small volumes. Slides may be removed to new four-well incubation trays by gently prying up the slide using the tips of the tweezers. Only lift the slides from the barcoded end or the end furthest from the protein content.
10. Serum is usually diluted between 1:150 and 1:500. It is helpful to prepare the diluted serum in advance during the blocking step by dispensing 5 mL of wash buffer for each sample into a 14 mL Falcon tube on ice and diluting the sera into these tubes. Cap each tube and mix by inverting the tubes several times. Care should be taken when adding the samples to the wells and during shaking so as not to cross contaminate the adjacent wells.
11. The pipette tip should be changed each time when aspirating the samples before the first wash step to avoid cross contamination. After this the same tip may be used to aspirate wash buffer. Do not tilt the trays to aspirate the samples from the wells as this could result in cross contamination of wells.
12. It is important to shield the secondary antibodies from light to prevent photodamage to the fluorescent labels. If the lights in the cold room cannot be turned off during the 90 min incubation step, the trays may be covered with tinfoil. Fluorescently labeled secondary antibodies to various Ig isotypes (e.g., IgG, IgM, and IgA) are commercially available. Also, Molecular Probes carries a number of kits for labeling secondary Ab's with Alexa Fluor dyes if there is no commercially available secondary Ab to a particular isotype or species of interest. It may be useful to include an additional secondary detection antibody

that has a different excitation and emission spectra than the first as a control. This antibody may recognize an epitope tag on the arrayed proteins to aid in data collection and analysis. For example, Alexa Fluor 647 anti-human IgG may be combined with Alexa Fluor 555 anti-GST if the proteins on the array were GST purified. Secondary detection antibodies are usually diluted 1:2,000 in wash buffer. Secondary antibodies must be prepared fresh each day. It is convenient to dilute the detection antibodies during **step 5** above. After dilution secondary Ab's should be kept in the dark.

13. It is not necessary to change pipette tips during this wash cycle.
14. This step improves the signal to noise ratio by removing background caused by compounds in the buffer (e.g., salts and detergent) and unbound secondary antibody. Arrays may be submerged two to four times each (be consistent in the treatment of each array). It is convenient and more consistent to submerge all of the arrays that have been probed in a batch at the same time by placing them into a slide rack and submerging the entire rack.
15. This step helps to remove H_2O that would otherwise dry on the array surface leaving areas of high background. The slide boxes should be lined with filter paper or Kimwipes to absorb excess moisture.
16. The overnight incubation is important to assure that there is no additional moisture on the arrays prior to scanning. The arrays may be placed at 4 °C after scanning for storage if it is important to rescan the arrays (Alexa Fluor dyes remain stable for over multiple scans).
17. Molecular devices' scanners are commonly used for scanning FAST/PATH based arrays. PMT settings should be determined by scanning one randomly selected array. Settings must be chosen such that the images display a dynamic range of high and low signals and the same settings should be used for scanning all of the arrays that will be compared. If two secondary antibodies were used, ratio data may also be collected. This is much like array co-genome hybridization (CGH) with nucleic acid arrays.
18. Normalization and statistical comparison methods that have been used for nucleotide based microarrays are often employed for analyzing protein arrays data. Common statistical methods of comparison include Mann–Whitney *U*-test, Student's *T*-test, and *M*-statistic, in order to determine which antigens are statistically differentially reactive between the two groups [2].

References

1. Ptacek J et al (2005) Global analysis of protein phosphorylation in yeast. Nature 438(7068): 679–684
2. Hudson M et al (2007) Identification of differentially expressed proteins in ovarian cancer using high-density protein microarrays. Proc Natl Acad Sci U S A 104(44):17494–17499
3. Beyer N et al (2012) Investigation of autoantibody profiles for cerebrospinal fluid biomarker discovery in patients with relapsing–remitting multiple sclerosis. J Neuroimmunol 242:26–32
4. Chen R et al (2012) Personalized omics profiling reveals dynamic molecular and medical phenotypes. Cell 148:1293–1307
5. Zhu H et al (2006) Severe acute respiratory syndrome diagnostics using a coronavirus protein array. Proc Natl Acad Sci U S A 103(11): 4011–4016
6. Scanlan M et al (2001) Humoral immunity to human breast cancer: antigen definition and quantitative analysis of mRNA expression. Cancer Immun 1:4
7. Qiu J et al (2011) Nucleic acid programmable protein array a just-in-time multiplexed protein expression and purification platform. Methods Enzymol 500:151–163
8. Carlsson A et al (2008) Serum proteome profiling of metastatic breast cancer using recombinant antibody microarrays. Eur J Cancer 44:472–480
9. Molina R et al (1998) c-erbB2 oncoprotein, CEA, and CA 15.3 in patients with breast cancer: prognostic value. Breast Cancer Res Treat 51:109–119
10. Levenson V (2007) Biomarkers for early detection of breast cancer: what, when, and where? Biochim Biophys Acta 1770:847–856
11. Disis M et al (1997) High-titer HER-2/neu protein-specific antibody can be detected in patients with early-stage breast cancer. J Clin Oncol 15(11):3363–3367
12. Cho W (2007) Contribution of oncoproteomics to cancer biomarker discovery. Mol Cancer 6(25):1–13
13. Zhu X et al (2006) ProCat: a data analysis approach for protein microarrays. Genome Biol 7:R110

Chapter 15

Interrogation of In Vivo Protein–Protein Interactions Using Transgenic Mouse Models and Stable Isotope Labeling

Anwesha Dey, Jiansheng Wu, and Donald S. Kirkpatrick

Abstract

Methods in mass spectrometry have evolved in recent years, facilitating proteomic analyses that were previously beyond the limits of the technology. Transgenic mouse models, coupled with mass spectrometry proteomics, have served as valuable platform for elucidating the in vivo function of individual genes and proteins. Here we discuss the methods we have recently employed to characterize protein–protein interactions and posttranslational modifications in tagged knock-in mouse models. These methods can be broadly applied to other systems for various applications in both basic and translational science.

Key words SILAC, Mass spectrometry, Flag tagged KI mice, In vivo proteomics

1 Introduction

Through advances in biochemistry, genetics, and molecular biology, the roles for thousands of genes in normal mammalian physiology and disease progression have been elucidated. Focused studies in patient groups have revealed numerous genetic lesions which directly cause, or increase the risk of acquiring, human disorders. Genetically engineered mouse models make it possible to investigate how these genes, and their encoded proteins, contribute to pathogenic processes in mammalian tissues when disrupted or overexpressed. To understand the function of individual proteins, cell culture systems have been invaluable in mapping physical interactions and functional connections. Ultimately, biochemical, biophysical, and structural studies serve to close the circle by elucidating the precise mechanisms of a given protein that malfunction in disease and may be exploited for the development of therapeutics.

Despite advances in the area of in vivo proteomics, significant challenges have historically hindered our ability to investigate the tissue specific functions of individual proteins in the whole organism. Variables such as age, metabolic state, and immunological

Narendra Wajapeyee (ed.), *Cancer Genomics and Proteomics: Methods and Protocols*, Methods in Molecular Biology, vol. 1176, DOI 10.1007/978-1-4939-0992-6_15,

status manifest as crosstalk between tissues and affect the protein composition of cells in ways that cannot be modeled in culture. The catalog of reagents for capturing and characterizing endogenously expressed proteins remains far from comprehensive, while the epitope tag-overexpression methods commonly used to perform protein–protein interaction studies in cultured cells remain difficult to translate in vivo. Even in cultured cells, the introduction of epitope tagged proteins can result in anomalous protein–protein interactions stemming from overexpression artifacts or the absence of functionally relevant interacting proteins within model systems [1]. Recently, transgenic mouse models have begun to bridge these gaps. Strategies that encode epitope tags into proteins via their endogenous genomic loci have been established for interrogating protein–protein interactions and posttranslational modifications in the context of normal tissue-specific expression patterns. Such approaches circumvent the need for bait-specific antibody reagents and permit establishment of systematic protocols for protein enrichment from their endogenous environment [2]. Recent innovations involving the Zinc-Finger Nucleases (ZFNs) [3], Transcription Activator-Like Effector Nucleases (TALENs) [4], and Clustered Regularly Interspaced Short Palindromic Repeats (CRISPR) [5–7] offer to extend these efforts to the scale, and with the precision required for systematic proteomic investigation.

From the proteomics perspective, a biological system can be defined by the abundances, states and interactions of its proteins. In this context, quantitative mass spectrometry methods have emerged as a valuable tool for investigating protein level differences between biological samples [8]. Stable isotope labeling methods originally established for cultured cells, have been extended to the labeling of higher organisms including mammals [9–14], and have even been employed in the clinical setting [15, 16]. Metabolic labeling is particularly effective for protocols involving enrichment of selected proteins and posttranslational modifications. As compared to chemical labeling, a primary benefit is that lysates from metabolically labeled samples may be mixed together early during experimental procedures in order to account for differences in sample handling [17]. In immunoprecipitation-mass spectrometry (IP-MS) analyses, samples with and without tagged bait proteins can be mixed at the protein level either prior to IP (i.e., purification after mixing; PAM), or prior to SDS-PAGE and proteolytic digestion (i.e., mixing after purification; MAP) [18] as a means of assessing proteins that interact stably and transiently with a given complex. By combining genetically engineered mouse models with optimized methods for protein immunoaffinity enrichment and mass spectrometry, we and others have begun to investigate the tissue specific effects of proteins [2, 19, 20]. Here, we discuss the methods employed during our recent investigation of protein–protein interactions and posttranslational modifications in transgenic mouse models (Fig. 1).

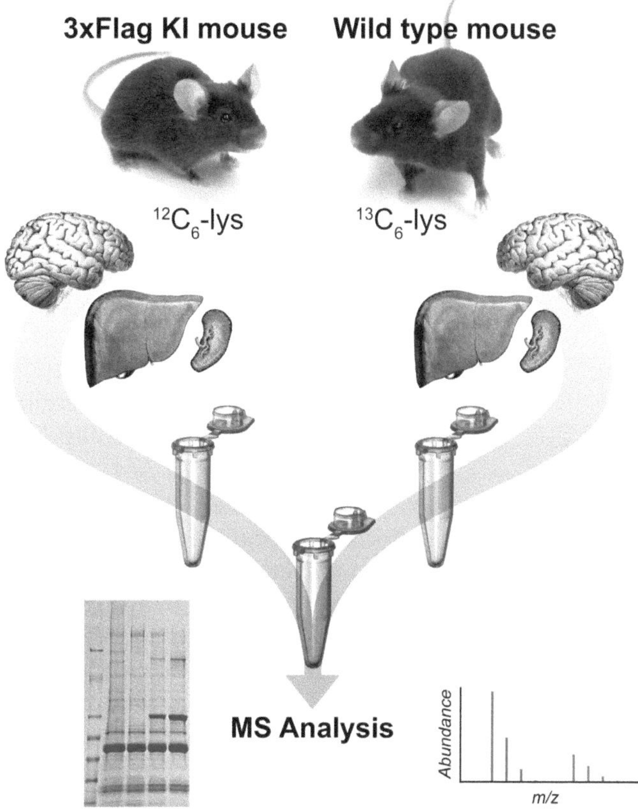

Fig. 1 Schematic for identification of protein–protein interactions using tissues from tagged knock-in (KI) mice. Tagged KI mice are fed regular chow, while wild type mice are fed labeled diet to make tissue lysates for flag immunoprecipitation in this case. Refer to text for more details

2 Materials

2.1 Flag Tagged KI Mice

Tagged knock-in mice are generated by homologous recombination as per standard established protocols for mouse generation. Using this approach, epitope tags are introduced into genes such that they remain under the control of the endogenous regulatory elements. In this way, epitope tagged proteins display normal levels and patterns of expression in tissues relative to untagged proteins, both in the basal state and upon perturbation. From the proteomics perspective, epitope tags that have proven to work well for immunoprecipitation in cell culture can likewise be used in vivo. While short peptide based tags are preferred, protein tags such as green fluorescent protein (GFP) also warrant consideration. Depending on the function of a protein, epitope tags may be inserted at either the N- or C-terminus, or in some cases, internally [19]. It is critical to ensure that the tag, even introduced under the control of endogenous regulatory elements, does not interfere with the function of the gene. In many cases, it is prudent to carry out cell-based studies in a standard model system prior to engineering epitope tagged mice for tissue specific investigation.

2.2 SILAC Mice

$^{13}C_6$-lysine-labeled wild type tissues of interest can be purchased from vendors including Cambridge Isotope Laboratories (Cambridge, MA). For projects involving multiple tissues, it can be more cost effective to isotopically label mice at an individual institution's with mouse facilities by feeding a defined diet in which individual amino acids are replaced by the isotopically labeled form. While most cell culture studies utilize a combination of isotopically labeled lysine and arginine, the costs associated with in vivo labeling have typically limited efforts to either a single amino acid or the SILAM approach [9, 11, 13, 14]. For amino acid labeling, lysine has typically been the amino acid of choice since it is essential in mammals and is not converted into other amino acids through the biosynthetic pathways [21]. Considering that trypsin and Lys-C proteases are commonly employed tools in MS proteomics and that approximately half of all tryptic peptides and nearly all Lys-C digested peptides carry a lysine residue to permit quantitation, this residue is ideal for most applications. For more details on generation of the SILAC labeled mouse, please refer to other chapters focusing on this topic [21].

2.3 Buffers

Flag IP lysis buffer contains 50 mM Tris–HCl, pH 7.4, 150 mM NaCl, 1 % Triton X-100, 1 mM EDTA. When stored at 4 °C Flag-IP, buffer is good for 1 year. Immediately prior to initiating the protocol, Complete Protease Inhibitors (Roche, Pleasanton, CA) are added to prepared Flag-IP lysis buffer. It will take several minutes to dissolve the tablet on a nutator mixer. After dissolving the protease Inhibitors, keep the Flag-IP lysis buffer on ice.

Flag-IP Wash buffer contains 50 mM Tris–HCl, pH 7.5, 150 mM NaCl and may be prepared ahead of time and stored at room temperature. Final elution of bait and interacting proteins is performed using 0.2 mg/ml 3×Flag peptide (Sigma) in Flag-IP elution buffer (50 mM Tris–HCl pH 7.4, 150 mM NaCl, and 1 mM EDTA).

2.4 Lysate Preparation

In this protocol, heavy (wild type) and light (Flag knock-in) mouse tissues are lysed separately (*see* **Note 1**). Following determination of protein concentration and before IP, lysates from light mouse and heavy mouse tissue are mixed in 1:1 ratio. Another slightly different approach involves performing IPs with heavy and light mouse lysate in different tubes and mixing the eluates before loading on SDS-PAGE gel. The benefits and drawbacks of each approach have been described in detail previously [18].

2.5 Mass Spectrometry

For LC/MS analysis, all solvents are prepared using Optima LC-MS grade solvents (ThermoFisher Scientific, Waltham, MA). Buffer A is comprised of 98 % H_2O–2 % acetonitrile–0.1 % formic acid, while Buffer B contains 98 % acetonitrile–2 % H_2O–0.1 % formic acid. Samples are resuspended in a buffer containing 2 % acetonitrile–0.1 % formic acid.

3 Methods

In this section, we describe detailed methods for immunoprecipitation of Flag-tagged bait proteins from transgenic mouse tissue for analysis of protein–protein interactions and posttranslational modifications by mass spectrometry. Isotopically labeled ($^{13}C_6$-lysine) mouse tissues were employed as spike-in standards to assess nonspecific interacting proteins captured during the immunoprecipitation process and understand tissue specific variability. Given the cost of isotopically labeled mouse tissues, initial optimization of the IP procedure can be done with unlabeled tissues alone. Major parameters to be optimized include amounts of mouse tissue lysate and anti-Flag M2 resin to be used in each IP. Pilot experiments can be performed at ~10 % scale and analyzed by silver staining relative to untagged wild type mice. Following optimization of buffers and immunoprecipitation conditions, IPs are performed in the presence of heavy and light tissues. As described below, the method has been employed in multiple efforts within our groups and proven applicable across numerous different tissues and bait proteins of interest.

1. Tissues from Flag-tagged knock-in transgenic mice and $^{13}C_6$-lysine-labeled wild type (Cambridge Isotope Laboratories) are collected during necropsy, flash frozen in liquid nitrogen and stored at −80 °C (*see* **Notes 2** and **3**). When possible, tissues should be cut into small pieces of 100–200 mg and stored in 1.5 ml tubes.
2. Transfer 1–2 ml of anti-Flag M2 resin (Sigma, St. Louis MO) into 10 ml Flag-IP lysis buffer and mix well. Allow the resin sit at 4 °C overnight. Immediately prior to initiating Flag-IP protocol, remove supernatant from the resin and resuspend the beads 1:1 in freshly prepared Flag-IP lysis buffer.
3. Weigh the tissues before use. For example, adult mouse brains weighs ~450 mg, spleens ~60–70 mg, and livers ~1,400 mg. For each 100 mg tissue to be used as input for IP, add 1 ml Flag-IP lysis buffer containing protease Inhibitors (*see* **Note 4**). Tissues from heavy mouse and light mouse are lysed separately.
4. Homogenization is performed with PowerGen 125 at speed setting 5. The first burst should break the tissue into very fine

particles in <1 min. After the first burst, allow the mixture sit on ice for 3–4 min, then repeat the homogenization for two additional times. The objective of homogenization is to get rid of visible particles in the solution. After three bursts, there should not be any visible particles in the mixture.

5. Allow tissue lysate to sit on ice for a total of 30–40 min with occasional mixing. Use 0.1–0.2 ml of lysis buffer to rinse the PowerGen tips and collect residual lysate. Transfer the mixture to 1.5 ml centrifuge tubes.
6. Centrifuge samples at 16,000 × *g* for 15 min using a benchtop centrifuge. For brain lysates, there are no lipid layers on top, but pellets of brain homogenate are quite large. For liver lysates, significant amounts of lipid are observed at the top, which can be removed by 0.45 μm spin filter (UFC30HV00; Millipore, MA, USA). This is important because excess lipids can prevent the lysate from easily passing through filter used in later steps.
7. Measure the protein concentration of lysates with Pierce Coomassie Plus Protein Assay (CAT#1856210). The concentration of brain lysate should be 8–10 mg/ml, liver lysate 17 mg/ml and spleen lysate 8–10 mg/ml. For each IP sample, between 2 and 20 mg of lysate should be used based on the results of optimization experiments. For Flag-BAP1 IPs performed from brain and spleen, we found that 5 mg total protein provided sufficient input material. Equal amounts of protein from a "light" Flag knock-in and "heavy" wild type lysate are mixed 1:1 at this step [20].
8. Add 3 μl of anti-Flag M2 resin to 200 μl lysate (8–10 mg/ml; *see* **Note 5**). To pipette 3 μl of M2 resin accurately, begin with 80 μl of Flag-IP lysis buffer and add 20 μl 50 % anti-Flag M2 slurry from the manufacturer's package. This volume represents the minimum amount of M2 resin that can be accurately pipetted with using a p200 pipette. Remove 30 μl of the resin-buffer mixture (equivalent to 3 μl bed volume M2 resin) and mix with lysate in Ultra-free spin filter. For this step, it is important to use a 5 μm, and not 0.45 μm size filter. The larger 5 μm filter has superior flow rate, while 0.45 μm has a tendency to get clogged by concentrated cell lysate. The Ultra-free spin filter can hold <0.4 ml lysate. For larger volume samples (e.g., >2 ml), a 15 ml falcon tube can be used instead (*see* **step 10b**).
9. Incubate the mixture at 4 °C on a nutator mixer for 1 h. No benefit is observed for longer incubation times.

10a. Spin Ultra-free spin filter containing M2 resin and lysate at 2,700 × *g* for 1 min at room temperature. Use vacuum to remove lysate solution from the bottom tube. A small portion (10–30 μl) of this lysate solution may be saved for use at the end of the protocol.

10b. If a 15 ml tube was not used in **step 8**, centrifuge the conical tube at 2,000 × *g* for 4 min. Use vacuum to remove most supernatant, leaving 0.3 ml solution at the bottom to prevent the loss of beads. Transfer the leftover solution and beads into a fresh Ultra-free tube and proceed as in **step 10a**.

11. First wash: add 0.33 ml Flag-IP lysis buffer to the beads. Incubate at room temperature for 5 min and then centrifuge at 2,700 × *g* for 1 min as in **step 10a**. The maximum volume that can safely be added to the Ultra-free spin filter is 0.33 ml.
12. Repeat **step 11** two additional times for a total of three washes.
13. To the beads, add 0.3 ml Flag-IP Wash buffer (50 mM Tris–HCl, pH 7.5, 150 mM NaCl). This buffer is stored at RT. Spin the tube at 2,700 × *g* for 1 min and remove filtrate carefully to prepare for the next elution step.
14. To elute proteins from the anti-Flag M2 resin, add 30 μl Flag-IP Elution Buffer containing 3×Flag peptide at a concentration of 0.2 mg/ml. The M2 resin is easily resuspended during the process of adding the elution buffer to the resin. This 30 μl of Flag-IP Elution Buffer volume is optimized for resin volumes up to 10 μl.
15. Place the suspension containing M2 resin and captured proteins in Flag-IP Elution Buffer on the nutator mixer and incubate for 30 min at room temperature.
16. Prior to collecting eluted proteins, replace the Ultra-free spin filter tube with a fresh tube. Centrifuge at 2,700 × *g* for 2 min to collect ~30 μl of protein eluate.
17. Add 7.5 μl of 5× SDS sample buffer containing 20 mM DTT (final concentration) to 30 μl protein eluate and boil the mixture for 3–5 min. Cool the sample to room temperature and briefly centrifuge to maximize recovery.
18. Load 30 μl of the boiled protein eluate onto a single lane of a 10- or 12-well, 1 mm SDS-PAGE gel (Invitrogen).
19. Samples for MS analysis are stained with Simply Blue Coomassie. Portions of the sample may also be utilized for western blotting or silver staining during method optimization (*see* **Note 6**).
20. On a separate lane of the gel, usually separated by at least one empty lane, input lysate mixtures are also resolved to permit evaluation of mixing ratios. The flow through lysate from **step 10** can be used for this purpose.
21. Samples for LC-MS/MS analysis are prepared followed previous descriptions [22, 23], with minor modifications. The gels are cut into 10–16 regions depending on the tissue of interest and digested with either trypsin or Lys-C protease. Digested peptides are extracted, dried and prepared for analysis in resuspension buffer.

22. Samples are injected onto a 0.1 × 100 mm column packed with 1.7 μm BEH-130 C18 using a NanoAcquity UPLC (Waters) and introduced to a LTQ-Orbitrap XL mass spectrometer (ThermoFisher Scientific) through an ADVANCE electrospray ionization source (Michrom-Bruker). Peptides are separated across a 35 min linear gradient of Buffer A and Buffer B as previously described [22], with a 60 min run time. The Orbitrap is operated in data dependent top8 mode, with a full MS scan collected in the FTMS at 60,000 resolution across a mass range from 375 to 1,600 m/z and MS/MS collected in the ion trap on the top eight most abundant precursor ions using a 30 s dynamic exclusion window (±20 ppm). Similar settings are also applicable to Velos-Orbitrap and Orbitrap Elite instruments. In the latter case, full MS scan times are reduced by approximately fourfold and data are acquired on the top 15 precursor ions from each full MS scan.
23. Raw spectral data are converted to mzXML file format and corrected monoisotopic masses corresponding to each selected precursor determined using mass measurements collected across the chromatographic peak [24]. MS/MS spectra are searched with Mascot [25] using a 50 ppm precursor ion tolerance against a concatenated target-decoy database containing murine proteins extracted from the Uniprot database (Version 2010_12) and common contaminants using full enzyme specificity (e.g., trypsin/LysC specific). Lysine ($^{13}C_6$) and oxidized methionine are considered as variable modifications. Peptide spectral matches (PSMs) are filtered using linear discriminant analysis to a 1 % peptide level false discovery rate on a per run basis. Area under curve (AUC) measurements corresponding to the intensities of "light" and "heavy" peptides can be determined using the VistaQuant algorithm and filtered using a heuristic score cutoff of 83 [26]. Redundant observations of chromatographic peak pairs are removed. For pairs where either the "light" or "heavy" chromatographic peak area remained undetermined after consolidation, missing area values are reported relative to the signal-to-noise ratio of the identified peak. After removing common laboratory contaminants (e.g., keratins, digestive enzymes), which by definition match only unmodified peptides with extreme ratios, the data from each tissue–bait pair are combined and normalized to the median ratio (log_2) of all peak observations where both "light" and "heavy" AUC values were observed. To determine the relative protein abundance between "light" and "heavy" samples, normalized AUC values are summed for each peptide passing the heuristic score filter. Proteins are mapped to genes based upon the Uniprot database and putative interactors selected based on at least eight observed AUC pairs and a normalized protein log_2 ratio >3 (light Flag-bait/heavy WT).

24. Given that anti-Flag antibodies recognize and can enrich for certain proteins in wild type and transgenic tissues alike, a subtraction strategy was employed for our analyses [20]. Immunoprecipitations with tissues from an unrelated knock-in 3×Flag tagged mice, provided a further control for specificity (*see* **Note** 7).
25. This SILAC approach with the subtraction strategy using the tagged mice allows identification of a smaller numbers of the most quantitative, endogenous and specific interactors of the bait protein of interest. These can then be validated by traditional immunoprecipitation, followed by western blot, if suitable antibodies for the identified interactors are available.

4 Notes

1. Determining the required amount of input lysate for each bait protein is important to the success of the tissue IP and must be optimized for each bait. For example, some poorly expressed proteins require up to 20 mg lysate. On the other hand, IPs for highly expressed proteins may be performed with as little as 2 mg of lysate. Silver stained gels from pilot IPs are often useful for determining how much lysate will be required for a successful mass spectrometry analysis. Abundant bait proteins and their binding partners can be easily detected by silver staining. Even for bait proteins that are weakly visible by silver staining, we obtain robust MS identifications for the bait protein and its associates. In some cases, such as for BAP1, bait bands were not clearly distinguishable via silver staining due to the effects of modifications such as glycosylation, but were still suitable for MS analysis.
2. It has been previously noted that SILAC-labeled mouse chow formulations contain high levels of sucrose [21] and could adversely impact results of studies related to metabolism. Our investigations noted differences in the baseline proteomes of tissues from labeled mice as compared to unlabeled counterparts. Hence, it may be important (more so for certain projects) to match diets between labeled and unlabeled mice to ensure that metabolic differences don't underlie differences in the baseline proteomes. This maybe can be difficult to do with purchased tissues and is more easily accounted for during in house labeling.
3. Cost of reagents is a significant consideration when embarking on in vivo metabolic labeling studies. Original work suggested that labeling period of 4–6 weeks, using either $^{13}C_6$-lysine or a complete ^{15}N-diet was effective for labeling proliferating tissues and those with high protein turnover, but nonetheless, did not

lead to complete labeling of proteins in tissues such as brain or heart [9, 10]. One possibility is that partial metabolic labeling may be sufficient for effective quantitation, as has been shown in plant biology [27]. Optimized protocols that enable carefully controlled labeling of mice, as is the case for tissue culture models, would be valuable in controlling cost by enabling short-term studies with pulse labeling.

4. High protein concentration has potential benefits and has proven helpful for detecting weaker protein–protein interactions from tissues. Using the lysis conditions described, the protein concentration in brain lysate is ~10 mg/ml. If less lysis buffer is used in this step, higher protein concentration (e.g., 15 mg/ml) can be achieved, but the extraction of proteins from tissues will be less efficient. However, higher concentration cell lysates inevitably increase nonspecific binding to the M2 resin. We consistently observe high levels of actin, tubulin and other proteins which displayed SILAC ratios near equivalence and were hence disregarded.
5. In many protocols, 20 μl or more anti-Flag M2 resin is suggested for IPs of 0.5–1 ml lysate as a means of minimizing loss of resin during the washing steps. However, we find that most nonspecific binding is caused by the interactions between anti-Flag resin and cellular proteins. According to the published product data sheet, 2.7 μg of M2 anti-Flag antibody is coupled to 1 μl of M2 resin. Therefore, even for highly expressed bait proteins, 3 μl of M2 resin is more than sufficient to fully deplete the available bait protein from the lysate and deliver the nanogram levels of protein required for detection by MS. Conversely, when using 20 μl of M2 resin for an IP, only a very small fraction of the 54 μg anti-Flag antibody actually binds to the bait proteins, leaving the majority available to capture proteins nonspecifically. According to our experience, we find that the background is proportional to the amount of M2 resin used for an IP and that it is essential to reduce the amount of antibody as much as possible. The use of the Ultra-free spin filters make it possible to handle as low as 3 μl beads without significant loss of beads. Likewise, for incubation time, where incubations longer than 1 h have consistently increased the level of nonspecific binding to the M2 resin in our analyses.
6. In certain cases, when using our brief and low stringency washing conditions, silver staining has not revealed differential bands between wild type versus 3×Flag tissues with the exception of the bait itself. When the bait is visible, it can be useful to proceed because the subtractive approach made possible by the use of isotopically labeled tissue can reveal specific and quantitative interacting proteins that are obscured on the gel by nonspecifically binding proteins.

7. As additional mice have come available, it has been valuable to perform analyses for multiple Flag-tagged bait proteins from tissues using this unified protocol. This approach, in conjunction with SILAC internal standard, makes it possible to account for proteins that bind nonspecifically during the IP procedure.

Acknowledgements

The authors would like to thank the following collaborators for applications of the in vivo proteomics technology and for helpful discussions and assistance at various stages of the project: Lilian Phu, Daisy Bustos, Corey Bakalarski, Haitao Zhu, Kim Newton, and Vishva Dixit.

References

1. Dunham WH, Mullin M, Gingras AC (2012) Affinity-purification coupled to mass spectrometry: basic principles and strategies. Proteomics 12(10):1576–1590
2. Schnutgen F et al (2011) Resources for proteomics in mouse embryonic stem cells. Nat Methods 8(2):103–104
3. Cui X et al (2011) Targeted integration in rat and mouse embryos with zinc-finger nucleases. Nat Biotechnol 29(1):64–67
4. Hockemeyer D et al (2011) Genetic engineering of human pluripotent cells using TALE nucleases. Nat Biotechnol 29(8):731–734
5. Jinek M et al (2013) RNA-programmed genome editing in human cells. eLife 2:e00471
6. Mali P et al (2013) RNA-guided human genome engineering via Cas9. Science 339(6121): 823–826
7. Wang H et al (2013) One-step generation of mice carrying mutations in multiple genes by CRISPR/Cas-mediated genome engineering. Cell 153(4):910–918
8. Bantscheff M, Lemeer S, Savitski MM, Kuster B (2012) Quantitative mass spectrometry in proteomics: critical review update from 2007 to the present. Anal Bioanal Chem 404(4): 939–965
9. Wu CC, MacCoss MJ, Howell KE, Matthews DE, Yates JR III (2004) Metabolic labeling of mammalian organisms with stable isotopes for quantitative proteomic analysis. Anal Chem 76(17):4951–4959
10. Kruger M et al (2008) SILAC mouse for quantitative proteomics uncovers kindlin-3 as an essential factor for red blood cell function. Cell 134(2):353–364
11. Rauniyar N, McClatchy DB, Yates JR III (2013) Stable isotope labeling of mammals (SILAM) for in vivo quantitative proteomic analysis. Methods 61(3):260–268
12. Krijgsveld J et al (2003) Metabolic labeling of C. elegans and D. melanogaster for quantitative proteomics. Nat Biotechnol 21(8): 927–931
13. Savas JN, Toyama BH, Xu T, Yates JR III, Hetzer MW (2012) Extremely long-lived nuclear pore proteins in the rat brain. Science 335(6071):942
14. Huttlin EL et al (2009) Discovery and validation of colonic tumor-associated proteins via metabolic labeling and stable isotopic dilution. Proc Natl Acad Sci U S A 106(40): 17235–17240
15. Bateman RJ et al (2006) Human amyloid-beta synthesis and clearance rates as measured in cerebrospinal fluid in vivo. Nat Med 12(7):856–861
16. Mawuenyega KG et al (2010) Decreased clearance of CNS beta-amyloid in Alzheimer's disease. Science 330(6012):1774
17. Ong SE et al (2002) Stable isotope labeling by amino acids in cell culture, SILAC, as a simple and accurate approach to expression proteomics. Mol Cell Proteomics 1(5):376–386
18. Wang X, Huang L (2008) Identifying dynamic interactors of protein complexes by quantitative mass spectrometry. Mol Cell Proteomics 7(1):46–57
19. Rao A, Richards TL, Simmons D, Zahniser NR, Sorkin A (2012) Epitope-tagged dopamine transporter knock-in mice reveal rapid endocytic trafficking and filopodia targeting of

the transporter in dopaminergic axons. FASEB J 26(5):1921–1933

20. Dey A et al (2012) Loss of the tumor suppressor BAP1 causes myeloid transformation. Science 337(6101):1541–1546
21. Zanivan S, Krueger M, Mann M (2012) In vivo quantitative proteomics: the SILAC mouse. Methods Mol Biol 757:435–450
22. Phu L et al (2011) Improved quantitative mass spectrometry methods for characterizing complex ubiquitin signals. Mol Cell Proteomics 10(5):M110 003756
23. Sheng Z et al (2012) Ser1292 autophosphorylation is an indicator of LRRK2 kinase activity and contributes to the cellular effects of PD mutations. Sci Transl Med 4(164): 164ra161
24. Haas W et al (2006) Optimization and use of peptide mass measurement accuracy in shotgun proteomics. Mol Cell Proteomics 5(7):1326–1337
25. Perkins DN, Pappin DJ, Creasy DM, Cottrell JS (1999) Probability-based protein identification by searching sequence databases using mass spectrometry data. Electrophoresis 20(18): 3551–3567
26. Bakalarski CE et al (2008) The impact of peptide abundance and dynamic range on stable-isotope-based quantitative proteomic analyses. J Proteome Res 7(11):4756–4765
27. Huttlin EL, Hegeman AD, Harms AC, Sussman MR (2007) Comparison of full versus partial metabolic labeling for quantitative proteomics analysis in Arabidopsis thaliana. Mol Cell Proteomics 6(5):860–881

Chapter 16

New Biophysical Methods to Study the Membrane Activity of Bcl-2 Proteins

Stephanie Bleicken and Ana J. García-Sáez

Abstract

The proteins of Bcl-2 family are key regulators of apoptosis. Many Bcl-2 proteins have the unique ability to switch between two possible conformations, soluble in the cytosol or associated to cellular membranes. Importantly, their membrane-inserted form is the main responsible for their apoptotic function. Unfortunately, there are only a limited number of methods available to study the membrane activity of these proteins. Here, we present a methodology to study protein binding to membranes and membrane permeabilization at the single vesicle level. It is based on purified proteins and giant unilamellar vesicles and involves directly visualization of the process with a confocal microscope. This approach allows for the characterization of the membrane activity of the Bcl-2 proteins (or of any other pore-forming molecule) with unprecedented detail.

Key words Bcl-2 proteins, Membranes, Giant unilamellar vesicles, Confocal microscopy, Pore formation, Protein membrane binding

1 Introduction

All living organisms are built up by a multitude of aqueous reaction compartments separated by hydrophobic lipid bilayers. During evolution, most proteins have been optimized to function either in aqueous or in lipophilic environments, and can be classified into soluble or membrane proteins accordingly. However, there are some proteins with the unique ability to switch between water-soluble and membrane-inserted conformations and can therefore have functions in both worlds.

Originally mainly toxins were identified to have this rare ability. Toxins can be found in all kingdoms of life and use a variety of mechanisms to kill their target cells. Some of them, like melittin, diphtheria toxin, sticholysins, or colicins, make pores into membranes [1] and thereby kill cells. For these proteins, the ability to switch from a soluble to a membrane inserted conformation in mandatory.

Narendra Wajapeyee (ed.), *Cancer Genomics and Proteomics: Methods and Protocols*, Methods in Molecular Biology, vol. 1176, DOI 10.1007/978-1-4939-0992-6_16, © Springer Science+Business Media New York 2014

Another fascinating family of proteins known to switch between soluble and membrane-inserted conformations are the Bcl-2 proteins. They are key regulators and executioners of the intrinsic apoptotic pathway and play a major role in a cell's decision to commit suicide [2, 3]. Mutations or changes in expression of different Bcl-2 proteins can lead to severe illnesses like cancer or neurogenerative diseases. They are also involved in the cellular responses to anticancer therapies. For these reasons, the Bcl-2 proteins have since long been acknowledged as important targets for drug design [4]. Unfortunately, many fundamental aspects of the molecular functions of the Bcl-2 proteins, like how signaling is orchestrated via their interaction network or the nature of their membrane inserted conformation, remain poorly understood [5, 6], which makes drug development a very difficult process. Due to the multitude of factors affecting the Bcl-2 proteins in the complex environment of the cell, many important questions cannot, or only with great difficulties, be studied in vivo. Alternatively, in vitro reconstituted systems have proven to be very useful tools to dissect the molecular mechanisms involved in Bcl-2 function (e.g., [7–12]).

Here, we present a set of methods that have been recently implemented to study the membrane activity of Bcl-2 proteins at the single vesicle level. The approach is based on the use of reduced systems composed of purified proteins and model membranes, concretely giant unilamellar vesicles (GUVs). GUVs have a comparable size to eukaryotic cells and can therefore be imaged by light microscopy [13, 14]. This allows for the study of membrane binding and permeabilization mechanisms of individual liposomes, which provides an unprecedented degree of detail as well as a direct observation of the process in the microscope. Further advantages of the GUVs system are the possibility to use well-defined and controllable lipid compositions and lipid bilayers visualized by fluorescent dyes.

The methods described here are designed to detect and quantify membrane binding, insertion, and permeabilization. We have successfully applied them to the Bcl-2 proteins before [7, 15]. However, they are not limited to the Bcl-2 proteins and can in general be applied to proteins that switch between soluble and membrane inserted conformations like toxins [16, 17], unconventional secreted proteins directly passing membranes [18], or others [19].

We explain four different experimental approaches to derive detailed information about protein membrane binding, membrane insertion and membrane permeabilization. The first experiment is designed to directly observe and compare binding to membranes with different lipid composition. Then, the mechanism of membrane permeabilization by the molecules of interest is characterized in detail by quantifying the degree of filling after a defined incubation time, the stability of the permeabilized state, and the kinetics of filling for the individual vesicles in the GUVs population.

The *filling-degree* experiment is designed to study whether membrane permeabilization follows a "graded" or "all-or-none" type. The *pore-stability* experiment allows to distinguish permanent from transient membrane permeabilization, and the *kinetic* experiments give information about the total permeabilized area and its progression over time (if the pore area is constant, growing, or shrinking over time).

2 Materials

1. Prepare the proteins of choice. The protein needs to be functional, and should be in the same buffer used for the experiment (here PBS). In the case of binding experiments, the protein needs be fluorescently labeled (e.g., with GFP or, preferably, organic dyes like Alexa488; here, we show cBid labeled with Alexa488 at Cys126) [20]. Usually, we use a final protein concentration between 5 and 100 nM (depending on the protein) in a total sample volume of 300 μl. To study membrane permeabilization we use purified cBid and Bax [21].
2. Determine the lipid composition to use and calculate the amounts of each lipid that should be mixed. Lipids solutions should be always stored in glass vials and handled with Hamilton syringes. Mix the lipids in chloroform, then dry the mixture completely under vacuum, and store it under nitrogen or argon atmosphere at −20 °C. For one experiment, 0.5 mg of the lipid mixture are needed. To study the membrane activity of Bcl-2 proteins, lipid mixtures mimicking the outer mitochondrial membrane (either a composition of 46 % egg PC L-α-phosphatidylcholine, 25 % egg PE L-α phosphatidylethanolamine, 11 % bovine liver PI L-α-phosphatidylinositol, 10 % 18:1 phosphatidylserine, 8 % cardiolipin—w/w; or a more simple mixture of 80 % egg-phosphatidylcholine and 20 % cardiolipin (mol/mol) [7, 9]) were prepared (*see* also **Note 1**).
3. Prepare buffer and sucrose solutions. We suggest using PBS, which contains physiological pH and salt concentration, as external buffer and preparing the GUVs in 300 mM sucrose solution. Other possibilities are valid. However, the solutions outside and inside the GUV samples need to be iso-osmolar.
4. For GUV preparation, a homemade GUV chamber made of Teflon and Platinum-wires (wire diameter: 0.4 mm; distance between the wires: 4 mm), a function generator and connecting cables are necessary (*see* Fig. 1a, b and **Note 2**).
5. Fluorescent, lipophylic dyes to label the membrane are used. Here experiments using DiI (1,1′-Dioctadecyl-3,3,3′,3′-Tetramethylindocarbocyanine Perchlorate) and DiD (1,1′-Dioctadecyl-3,3,3′,3′-Tetramethylindodicarbocyanine

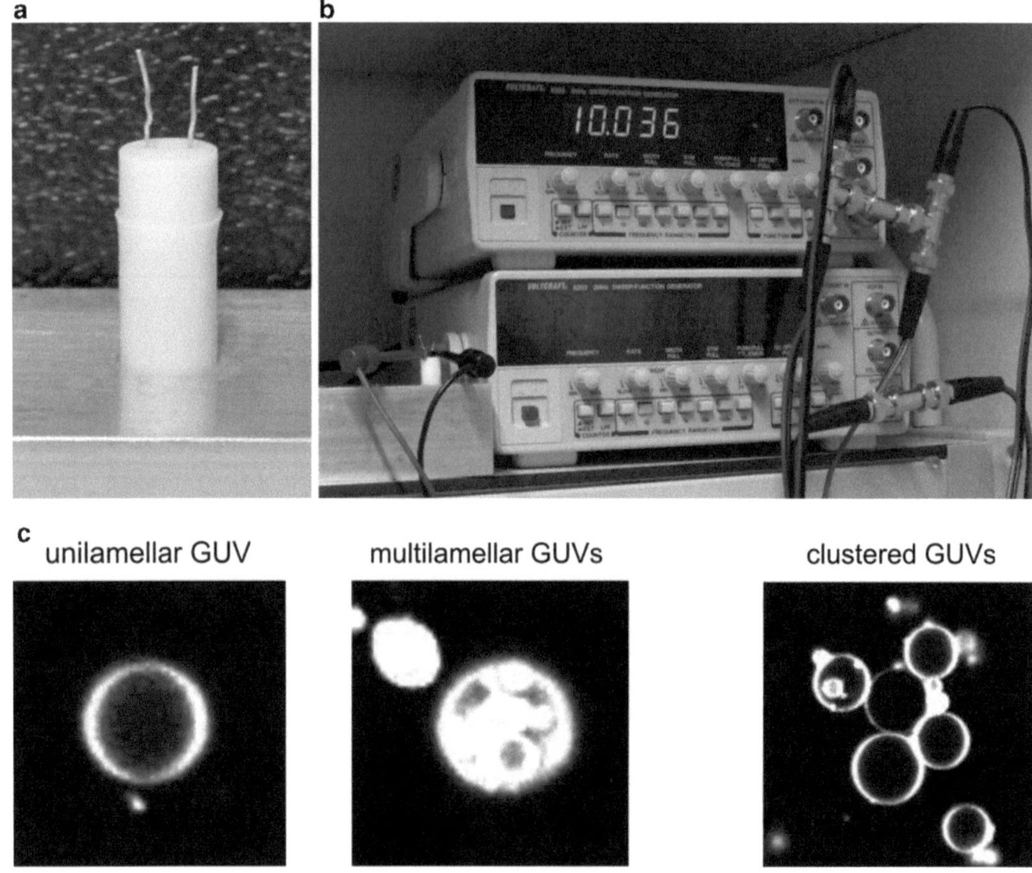

Fig. 1 Technical equipment needed for GUV electro-formation. (**a**) GUV electro-formation chamber. (**b**) Function generator with connecting cables and GUV chamber. (**c**) Example pictures of GUVs

Perchlorate) are shown. However DiO (3,3′-Dioctadecy loxacarbocyanine Perchlorate) or fluorescently labeled lipids are possible as well.

6. Buy/prepare suitable observation chambers to work in a confocal microscope (e.g., 8-well Labtek chambers; NUNC).
7. Prepare a 1 mg/ml Casein solution in PBS.
8. Buy/prepare fluorescent size markers of your choice. Here we use free Alexa488 and Alexa633 dyes (approx. 1 kDa). Fluorescently labeled dextran's or fluorescent proteins are possible as well and can be used to estimate the size of the individual pores in the GUVs [7].
9. Data collection proceeds in a confocal microscope. It needs to be equipped with laser lines and detectors suitable for the used fluorescent dyes and a water objective. The function "title scan" to visualize a larger sample area is very helpful. All experiments presented here were performed on an LSM710 microscope with

a C-Apochromat 40× 1.2 water immersion objective (Zeiss). Excitation light came from Ar-ion (488 nm), HeNe (561 nm) or HeNe lasers (633 nm). A spectral beam guide was used to separate the emitted photons from the different fluorophores.

10. For manual data analysis Image J was used (http://rsbweb.nih.gov/ij/). In the case of binding experiments, the plugin radial profile is required (http://rsbweb.nih.gov/ij/plugins/radial-profile.html).
11. For semi-automated data analysis our homemade "GUV detector" program can be used (*see* **Note 3**; please contact the authors for a free copy of the software).

3 Methods

3.1 GUV Preparation

1. Disolve 0.5 mg lipid mixture in a total volume of 0.5 ml chloroform (1 mg/ml). Add the lipidic dye to the lipid mixture in chloroform. Here, we used DiO (<0.05 %; excited with the 488 nm laser) to label the membrane for protein binding experiments, degree of filling experiments, and kinetic experiments. For pore stability experiments DiI is used (<0.05 %; excited with the 561 nm laser; *see* also **Note 4**). Make sure to use a lipophilic dye excited by a different laser than the size marker or the protein to analyze.
2. Calculate how many samples and therefore electro-formation reactions are needed for the experiment. In our case, one GUV formation chamber contains 350 μl sucrose solution, and we add 70 μl of GUV solution per well. Thus one reaction has enough GUVs for 4–5 wells.
3. Spread 2.5 μl of the lipid mixture on each platinum wire and let it dry (best 2 h under vacuum). Fill each electro-formation chamber with 350 μl of 300 mM sucrose solution. Close the chamber. More than 5 mm of each wire should be in contact with the solution. Apply 10 Hz and 1.5 V for 2 h and then reduce to 2 Hz for 30 min to detach the vesicles from the wires. Afterwards the GUVs are ready for the experiment (*see* also **Note 5**).

3.2 Preparation of the Observation Chamber

1. Calculate how many samples you need for the experiments and block the glass surface of the observation chambers by incubation with the casein solution (1 h at RT) to minimize unspecific protein/lipid binding during the experiment.
2. Remove the casein solution and wash the observation chambers thoroughly with Millipore water to remove free casein.
3. Test experiments: The conditions described here were optimized for the microscope described in the materials part.

However, different confocal microscopes have different sensitivities. Therefore, the experimental setup and eventually the amount of the fluorescent dyes should to be optimized for the used system. We suggest the implementation of test experiments. Suitable test samples are described in the experimental part described for each experiment. As a guideline, we usually measure with a pinhole aperture of 1–2.5 AU and a laser power of 0.5–5 %, and acquire an average of 2–4 scans at a medium scan speed. However the settings can vary between different systems.

4. The protocols of all four experiments are divided into experimental and data analysis sections. They are schematically drawn in Fig. 2. Each experimental part contains a "test sample" to optimize the microscopic setup. In the data analysis section the following nomenclature is used: Membrane channel: Channel of the membrane dye; Protein Channel: Channel of the dye attached to the protein; Size marker channel: Channel of dye attached to or used as size marker. Images and data plots of example experiments are shown in Fig. 3 (*see* **Note 6**).

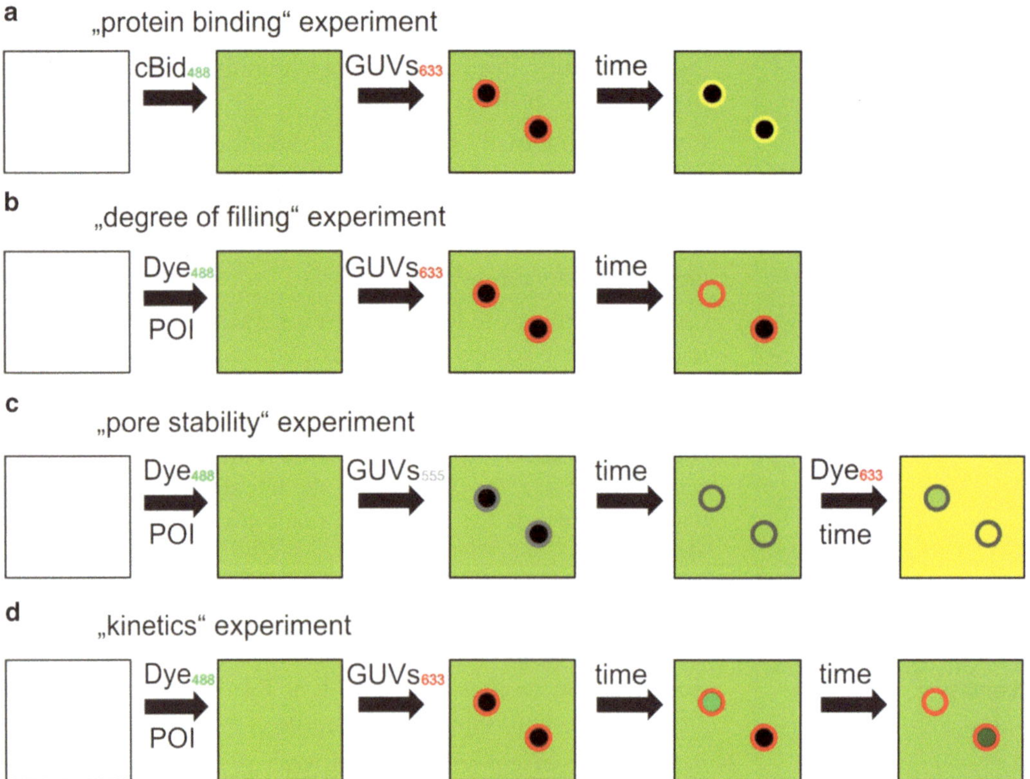

Fig. 2 Schematic drawing of the four different experimental setups. (**a**) Protein binding. (**b**) Degree of filling. (**c**) Pore stability. (**d**) Kinetics. *POI* protein of interest

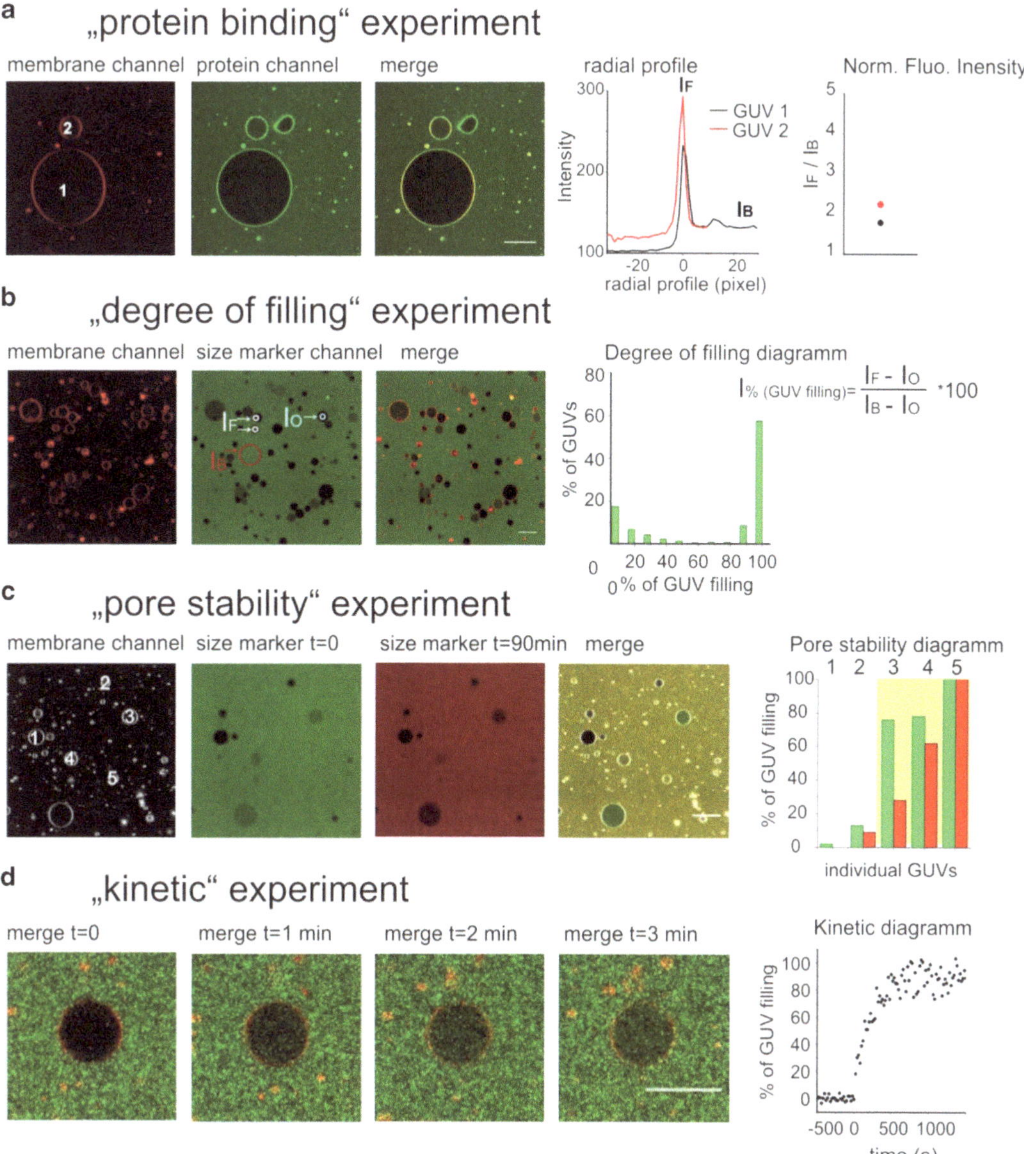

Fig. 3 Examples of images from the experiments and their corresponding quantitative analysis examples. (**a**) Protein binding experiment. *Left:* Example images. *Right:* radial profiles of the two GUVs in the images and intensity ration after normalization. (**b**) Degree of filling experiment. *Left:* Example images. Background determination (I_B). *Red arrow* and *circle* indicate a suitable area; Offset determination (I_O) *blue arrow* and *circle* indicate a suitable area; and GUV filling (I_F) *white arrows* and *circles* indicate areas. *Right:* Graph of analyzed data and formula used in the calculations. (**c**) Pore stability experiment. *Left:* Example images. *Right:* filling of individual GUVs for both dyes. (**d**) Kinetic experiment. *Left:* Example images. *Right:* filling of one individual GUVs over time. Scale bar: 40 μm

3.3 Protein Binding to Membranes

3.3.1 Experiment (Schematically Drawn in Fig. 2a)

1. For protein binding experiments dilute the fluorescently labeled protein of choice at the desired concentration with buffer. The volume of the mixture should be 230 μl and the protein concentration calculated for a final 300 μl total volume. Here, we use 20 nM cBid labeled with Alexa 488. Mix the sample rigorously.
2. In a second well prepare a negative control sample containing 230 μl buffer. This control is important to detect and remove potential channel cross talk during the setting up of the confocal microscope.
3. Transfer 70 μl DiD stained GUVs by using a pipette with a cut tip (the inner diameter should be at least 1 mm). Carefully mix the GUV suspension in the GUV chamber by pipetting slowly up and down and then carefully take 70 μl suspension and add them into each observation chamber.
4. Incubate GUVs and proteins together before imaging in the confocal microscope. Here, we incubate for 60 min at RT. Note: Incubate at least 5–15 min to allow the GUVs to sink to the bottom of the chamber.
5. Test samples: For this experiment three test samples are prepared: (a) GUVs in buffer, (b) protein of choice in buffer, (c) all three components together. Optimize the microscope setup so that the membrane channel shows no or minimal cross talk with the protein channel. This step is critical for the experiment as the cross talk can be otherwise detected as protein binding.
6. Image your sample in the confocal microscope. Make sure that you image enough unilamellar vesicles to do statistics (*see* also **Notes 6** and 7).

3.3.2 Data Analysis

1. Data analysis can be performed manually in Image J or semi-automatically with the program "GUV detector" (*see* **Note 3**). Here, the manual analysis strategy is described.
2. For manual analysis Image J with the radial profile plug-in installed are necessary.
3. Open the first image in Image J. The two channels should be visible as a stack. In case your images are very dim you may increase the brightness for visualization, but you must not apply these changes to avoid changing the raw data. Example images are shown in Fig. 3a.
4. Choose a unilamellar GUV of interest and the protein channel, which is in our case the 488 nm channel. Use only unilamellar GUVs (*see* Fig. 1c).
5. Use the circular mask to overlay it with the GUV. It should have exactly the same size and position. To achieve this you may zoom in to enlarge the GUV. In case your binding is very weak you can do the positioning in the membrane channel, but

make sure that you switch to the protein channel, before you read out the intensities. Run the plug in radial profile. A new window pops up. It shows the xy position of the circle in the image and the radius of the circle. Enlarge the GUV radius by 20 % and press ok. The window radial profile plot pops up. Press list to get the xy-data to plot the radial profile. An example of the results is shown in Fig. 3a.

6. Repeat the step for many GUVs to achieve a good statistical analysis. Redo the process with several images from independent experiments.
7. For plotting the data as shown in Fig. 3a (right diagram), the data were normalized by measuring the maximum intensity at the GUV rim (I_F) and divide it by the background intensity (I_B).

3.4 Degree of Filling

3.4.1 Experiment (Schematically Drawn in Fig. 2b)

1. Test sample: Use a mixture of 225 μl buffer, 2–6 μl Alexa 488 (in water or buffer) as external dye (final concentration: 50–500 nM), and 70 μl DiD stained GUVs. In this experiment the concentration of the external dye and the microscope settings should be optimized. Incubate for 5–15 min before imaging to allow the GUVs to settle.
2. Calculate the number of reactions you need for the experiment. One negative (buffer) control is needed. In case you have one protein to induce permeabilization and want to test only one concentration, only two chambers are needed. For example, in the experiment shown here, we used a mixture of two proteins (cBid and Bax) to induce permeabilization. Therefore we prepared four reactions in total: (1) control, (2) cBid alone, (3) Bax alone, and (4) cBid + Bax.
3. Mix the external dye and the non-labeled protein/s of choice with the buffer, so that the mixture has a volume of 230 μl. Here, we used 10 nM cBid/20 nM Bax. Mix the sample rigorously. There are variations of the method possible (*see* **Note 8**).
4. Cut a pipette tip, so that the inner diameter is at least 1 mm. Carefully mix the sucrose GUV suspension in the GUV chamber with the tip and then carefully transfer 70 μl solution into each observation chamber. Mix very gently with the cut tip to have a reaction mix as homogeneous as possible without disrupting the vesicles.
5. As imaging takes time, do not add the GUVs suspension to all wells at the same time, but do it successively to use always the same incubation time.
6. Incubate GUVs and proteins before imaging in the confocal microscope. Here, we incubate for 60 min at RT. The timing is critical in this experiment, visualization should be carried out in or close to equilibrium.
7. Image enough GUVs per reaction to do statistics (we normally analyze 200–600 GUVs per reaction; *see* **Note 7**).

3.4.2 Data Analysis

1. Data analysis can be performed manually in Image J or semi-automatically with the program "GUV detector" (*see* **Note 3**). Here, the manual analysis strategy is described.
2. Open the first image in Image J. The two channels should be visible as a stack. In case your images are very dim you may increase the brightness in for visualization, but do not apply these changes to avoid changing your raw data. Example images of the different channels are shown in Fig. 3b.
3. Background determination (I_B): Open the size marker channel (here Alexa488) and measure the background intensity (I_B) of the dye in the image by putting the circular mask over a representative GUV-free region of the image and apply the function measure (menu: analyze: measure; example regions are highlighted in Fig. 3b). The mean value should represent the background intensity. Repeat the process at least twice to make sure to work with a background of representative intensity. In case the background is not homogenous the image needs to be subdivided into regions with a homogenous background. Each region needs to be treated like an individual image.
4. Offset determination (I_O): Open the size marker channel and choose the two GUVs, which seem to have the lowest intensity inside. Place the circular mask in the GUVs so that the circle is clearly smaller than the GUV itself and has no overlay with the GUV rim. Read out the intensity inside the GUV (menu: analyze: measure; example regions are highlighted in Fig. 3b). The mean value should represent the intensity offset (I_O) of this image.
5. Degree of filling of individual GUVs (I_F): Choose a GUV of interest in the membrane channel and place the circular mask in the GUV so that the circle has no overlay with the GUV Rim. Switch the channel and read out the intensity inside the GUV (menu: analyze: measure; example regions are highlighted in Fig. 3b). The mean value should represent the intensity inside the GUV (I_F). Use only unilamellar GUVs (*see* Fig. 1c).
6. Repeat the process for all unilamellar GUVs of the picture/region and all images from independent experiments.
7. For plotting the data as shown in Fig. 3b right side calculate the degree of filling of each individual GUV by applying the formula drawn in Fig. 3b. Each GUV should have a corresponding $I_{\%}$ value.

3.5 Pore Stability

3.5.1 Experiment (Schematically Drawn in Fig. 2c)

1. For the test experiment use a mixture of 220 μl buffer, 2–6 μl Alexa488, 2–6 μl Alexa633 (in water or buffer; final concentration of each 50–500 nM) and 70 μl suspension of DiI-stained GUVs. In this experiment the concentration of the free dyes and the microscope settings should be optimized,

paying special attention to avoid cross talk. Incubate for 5–15 min before imaging to allow the GUVs to sink to the bottom of the well.

2. Calculate the number of observation chambers you need for the experiment. More information is given in the section Degree of filling.
3. Add the Alexa488 dye and the non-labeled protein/s of choice at a final concentration of 10–100 nM into the wells. Here, we use 10 nM cBid/20 nM Bax. Mix the sample rigorously and fill each well to a total volume of 230 μl.
4. Cut a pipette tip, so that the inner diameter is at least 1 mm. Carefully mix the sucrose GUV solution in the GUV chamber with the tip and then carefully transfer 70 μl suspension into each used reaction chamber. Mix very gently with the cut tip or do not mix at all as GUVs are very fragile.
5. Note that imaging takes time. Do not add the GUV solution to all chambers at the same time. Do it successively to have always the same incubation time.
6. Incubate GUVs and proteins. In this experiment, we incubate for 90 min at RT. Note that the timeframe is critical for the experiment, as discussed in more detail in **Note 8** and **9**.
7. Add the Alexa 633 dye diluted to a volume of 20–30 μl to each sample well and incubate for another 10–30 min to allow for dye equilibration. Do not mix or only very gently mix with a cut pipette tip to avoid GUV destruction.
8. Image 200–600 GUVs per reaction well to do statistics (*see* **Note 7**).

3.5.2 Data Analysis

1. Data analysis can be performed manually in Image J or semi-automatically with the program "GUV detector" (*see* **Note 3**). Here, the manual analysis strategy is described.
2. Open the first image in Image J. The three channels should be visible as a stack. In case your images are very dim you may increase the brightness for visualization, but do not apply these changes to avoid changing the raw data. Example images of the different channels are shown in Fig. 3c.
3. Background determination (I_B): Open the channels of both size markers (here Alexa488 and Alexa633) and choose areas without GUVs to measure the background intensity (I_B) of the dye in each image. For this purpose, use the circular mask and place it over a region without GUVs and apply the function measure (menu: analyze: measure; exemplary suitable regions are shown in Fig. 3b) to each of the two channels for the size markers. The mean value should represent the background intensity. Repeat the process at least twice to make sure to work with representative intensity values and homogenous background.

In case the background is not homogenous the image needs to be subdivided into regions with a homogenous background. Each region needs to be treated like an individual image.

4. Offset determination (I_O): Open the images corresponding to the channels of the size markers and choose two GUVs, which seem to have the lowest intensity inside in both channels. Place the circular mask in the GUVs so that the circle is clearly smaller as the GUV itself and has no overlap with the GUV rim. Read out the intensity inside the GUV (menu: analyze: measure) in each channel. The mean value represents the intensity offset (I_O) of each size marker in that picture.
5. Percentage of filling of individual GUVs (I_F) in the two channels: Choose a unilamellar GUV of interest in the membrane channel and place the circular mask in this GUV so that the circle has no overlap with the vesicle rim. Switch the channel and read out the intensity inside the GUV (menu: analyze: measure) for both channels. The mean value represents the intensity inside the GUV (I_F) in each color. Use only unilamellar GUVs (*see* Fig. 1c).
6. Repeat the process for each new image or region.
7. For plotting the data as shown in Fig. 3c right, calculate the degree of filling of each individual GUV each of the both size marker channels by applying the formula in Fig. 3b.
8. Now for each GUV the two values $I_{\%\ Alexa488}$ and $I_{\%\ Alexa633}$ should be calculated. Transfer the data into two columns of you data analysis program of choice for plotting.
9. Sort the column of the dye added second here Alexa633 from the smallest to the largest numbers. You can plot the data as shown for 5 exemplary GUVs in Fig. 3c (right; GUVs highlighted in the image on the left). Make sure that during sorting $I_{\%\ Alexa488}$ and $I_{\%\ Alexa633}$ corresponding to one GUV are still in one row.
10. This plot contains information about several processes. First, it shows how many GUVs are permeabilized. To classify them, set a threshold. We usually define that all GUVs with >50 % filling with the size marker added at t0 (here Alexa488) are considered as permeabilized. In the example in Fig. 3c, 60 % of the GUVs are permeabilized (3 of 5). For clarity, it is often recommendable to plot only the permeabilized GUVs. Second, the plot gives information about the stability of the permeabilized state of the membrane, which is related to the transient or stable nature of the pores. Stable pores allow the passage of both dyes (GUV 4 and 5 in the example in Fig. 3c show stable pores), while transient pores close before the second dye is added (as in the case of GUV 3). Thus, in the example of Fig. 3c, 2 out of 3 GUVs show stable pores and 1 out of 3 a transient pore. Additional information can be found in **Note 8** and **9**.

3.6 Kinetics of GUV Permeabilization

3.6.1 Experiment (Schematically Drawn in Fig. 2d)

1. For the test experiment use a mixture of 225 μl buffer, 2–6 μl Alexa488 (in water or buffer, final concentration 50–500 nM) and 70 μl suspension of GUVs stained with DiD. In this experiment the concentration of the free dye and the microscope settings should be optimized. Incubate for at least 30 min before imaging to allow the GUVs to sink to the bottom of the well and stop diffusing.
2. In this case only one reaction well per experiment can be imaged, as it is a time lapse experiment.
3. Add the free dye and the non-labeled protein/s of choice at a final concentration of 10–100 nM into the wells. Here, we use 10 nM cBid/20 nM Bax. Mix the sample rigorously and fill each well to a total volume of 230 μl. There are variations of the method possible (*see* **Note 7**).
4. Transfer the chamber to the confocal microscope, which should be ready for use.
5. Cut a pipette tip so that the inner diameter is at least 1 mm. Carefully mix the sucrose GUV suspension in the GUV chamber with the tip and then carefully transfer 70 μl suspension into the observation chamber. Do not mix to avoid unnecessary fluxes.
6. Set up the microscope to perform a time lapse experiment. We recommend to choose a region in the sample with several GUVs in the field of view (>20). We normally acquire one image every 10–30 s for a total time period of 60–120 min. However, the conditions need to be optimized for the used protein/membrane ensemble (*see* also **Note 7**).

3.6.2 Data Analysis

1. Data analysis can be performed manually in Image J or semi-automatically with the program "GUV detector" (*see* **Note 3**). Here, the manual analysis strategy is described.
2. Open the first image in Image J. The two channels should be visible as a stack. In case your images are very dim you may increase the brightness for visualization, but do not apply these changes to avoid changing the raw data. Example images of the different channels are shown in Fig. 3d.
3. Background determination (I_B): Open the size marker channel (here Alexa488) and measure the background intensity (I_B) of the dye in the image by putting the circular mask over a representative region free of GUVs in the image. Check whether this area is suitable in all images of the stack. Apply the function plot *z*-axis profile (menu: image: stack: Plot *z*-axis profile). The mean value in each image of the stack represents the background intensity. Repeat the process at least twice to make sure you work with representative intensity values and homogenous background. In case the background is not

homogenous the image needs to be subdivided into regions with a homogenous background. Each region needs to be treated like an individual image.

4. Offset determination (I_O): Open the size marker channel and choose the two GUVs with the lowest intensity inside. Place the circular mask in the GUVs so that the circle is clearly smaller than the vesicle and has no overlap with the GUV rim. Read out the intensity inside the GUV (menu: analyze: measure). Check whether this value is valid for the whole stack. From our experience, this is usually the case. Otherwise use the function plot *z*-axis profile to determine I_O for each image of the stack.
5. Determine the degree of filling of all individual GUVs (I_F) at each picture of the stack: Choose a unilamellar GUV of interest in the membrane channel and place the circular mask in this GUV so that the circle has no overlap with the vesicle rim. Note, that often new GUVs appear in the field of view during the experiment. For this reason, it usually helps to look for suitable GUVs after time zero. Check whether the GUV moves during the time of the stack or not. In case the GUV is not moving, switch to the size marker channel and read out I_F for each image of the stack by applying plot *z*-axis profile. The mean values represent the intensity inside the GUV (I_F) at each time point. For GUVs that move, the position of the circular mask needs to be readjusted and each image of the stack needs to be measured independently with the measure function. Use only unilamellar GUVs (*see* Fig. 1c).
6. Repeat the process of I_F determination for many GUVs.
7. Repeat the process for each image or region.
8. For plotting the data as shown in Fig. 3d (right side), calculate the degree of filling of each individual GUV at each time point by applying the formula in Fig. 3b. Afterwards, each GUV at each time point should have a corresponding $I_{\%}$ value, which can be plotted as shown in Fig. 3d. More information can be extracted by fitting the curve with appropriate models (*see* **Note 10**).

4 Notes

1. As the experiments consume only very low amounts of lipid, we usually prepare higher amounts of the desired lipid mixtures, aliquot them in 0.5 mg portions and store them under Nitrogen or Argon gas at –20 °C for several weeks.
2. A simple electro formation chamber can be built of a 0.5 or 1.5 ml reaction tube and platinum wires.
3. In our group, a program for the automated detection and analysis of giant vesicles on digital microscopy images was

developed [22]. The tool is suited to analyze all experiments presented here and it decreased the analysis time by about 70–90 %. Please contact the authors to receive a free version of the program.

4. To increase the brightness of the vesicles, one might be tempted to increase the concentration of lipidic dye added to the lipid solution in chloroform. Note that this can influence the bilayer properties.
5. The duration of GUV electro-formation and the temperature that should be used vary between lipid mixtures. Prepare the GUVs above the transition temperature of the lipid mixture and test the duration of the process for that lipid mixture. When GUVs are formed at temperatures above RT, let them slowly cool down before use. The electro-formation method for GUVs was originally described in [23, 24] and recently reviewed in [14]. For the shown experimental conditions, it is critical to produce mainly unilamellar GUVs (*see* Fig. 1c). Moreover, it is mandatory to analyze only unilamellar GUVs. The electro-formation technique may produce multilamellar GUVs, depending on the lipid composition, but to produce reliable data they should be a minority. Protein binds to all kinds of GUVs and when most of the GUVs cannot be analyzed, this produces noise or even wrong results. One should also be aware of vesicle clustering (*see* Fig. 1c). Some proteins induce vesicle clustering, however; observations of clustered GUVs in absence of proteins may indicate problems in GUVs preparation (e.g., contaminated lipid or sucrose solutions).
6. Make sure that the confocal images a not taken at a position very close to the glass. Due to the possible binding of proteins or dyes to the glass high fluorescence signals can be observed near to the glass. Moreover, artificial vesicle structures due to glass binding can occur in that area. Make sure that you show representative areas. Therefore, it is recommendable to take pictures at different distances from the glass and decide after examining all samples, which distance works best. Make sure that you work with similar distances in all samples.
7. A variation of the degree of filling and the kinetics methods is to work with big and small size markers in a three color experiment. In this way, additional information about the pore size can be obtained. For example when a certain pore size is expected, two size markers can be chosen of which one is smaller than the expected pore size while the other is bigger. To avoid cross talk problems, it is recommended to use size markers with excitation maxima corresponding to the shortest and the longest wavelength laser lines, while the membrane dye corresponds to the laser line in between.

8. Pore stability experiment: It is critical to optimize the experimental settings so that the second dye is added under conditions close to equilibrium. New pores opening after adding the second dye are problematic, because they make the data noisy.
9. Pore stability experiment: The repeated opening and closure of pores in GUV over a long time period cannot be distinguished from a pore that remains continuously open during the experiment. Both scenarios lead to the same situation.
10. Kinetic experiment: the initial and relaxed total permeabilized area of each GUV, as well as the relaxation time, can be estimated by mathematical fitting to the data. For more information see [7, 15].

References

1. Iacovache I, van der Goot FG, Pernot L (2008) Pore formation: an ancient yet complex form of attack. Biochim Biophys Acta 1778:1611–1623
2. Lindsay J, Esposti MD, Gilmore AP (2011) Bcl-2 proteins and mitochondria – specificity in membrane targeting for death. Biochim Biophys Acta 1813:532–539
3. Westphal D, Dewson G, Czabotar PE, Kluck RM (2011) Molecular biology of Bax and Bak activation and action. Biochim Biophys Acta 1813:521–531
4. Leber B, Geng F, Kale J, Andrews DW (2010) Drugs targeting Bcl-2 family members as an emerging strategy in cancer. Expert Rev Mol Med 12:e28
5. Garcia-Saez AJ (2012) The secrets of the Bcl-2 family. Cell Death Differ 19:1733–1740
6. Leber B, Lin J, Andrews DW (2010) Still embedded together binding to membranes regulates Bcl-2 protein interactions. Oncogene 29:5221–5230
7. Bleicken S, Wagner C, García-Sáez Ana J (2013) Mechanistic differences in the membrane activity of Bax and Bcl-xL correlate with their opposing roles in apoptosis. Biophys J 104:421–431
8. Landeta O, Landajuela A, Gil D, Taneva S, DiPrimo C et al (2011) Reconstitution of proapoptotic BAK function in liposomes reveals a dual role for mitochondrial lipids in the BAK-driven membrane permeabilization process. J Biol Chem 286:8213–8230
9. Lovell JF, Billen LP, Bindner S, Shamas-Din A, Fradin C et al (2008) Membrane binding by tBid initiates an ordered series of events culminating in membrane permeabilization by Bax. Cell 135:1074–1084
10. Czabotar Peter E, Westphal D, Dewson G, Ma S, Hockings C et al (2013) Bax crystal structures reveal how BH3 domains activate bax and nucleate its oligomerization to induce apoptosis. Cell 152:519–531
11. Suzuki M, Youle RJ, Tjandra N (2000) Structure of Bax: coregulation of dimer formation and intracellular localization. Cell 103: 645–654
12. Montessuit S, Somasekharan SP, Terrones O, Lucken-Ardjomande S, Herzig S et al (2010) Membrane remodeling induced by the dynamin-related protein Drp1 stimulates Bax oligomerization. Cell 142:889–901
13. Sezgin E, Schwille P (2012) Model membrane platforms to study protein-membrane interactions. Mol Membr Biol 29:144–154
14. Walde P, Cosentino K, Engel H, Stano P (2010) Giant vesicles: preparations and applications. ChemBioChem 11:848–865
15. Fuertes G, Garcia-Saez AJ, Esteban-Martin S, Gimenez D, Sanchez-Munoz OL et al (2010) Pores formed by Bax alpha 5 relax to a smaller size and keep at equilibrium. Biophys J 99: 2917–2925
16. Apellaniz B, Garcia-Saez AJ, Huarte N, Kunert R, Vorauer-Uhl K et al (2010) Confocal microscopy of giant vesicles supports the absence of HIV-1 neutralizing 2F5 antibody reactivity to plasma membrane phospholipids. FEBS Lett 584:1591–1596
17. Schon P, Garcia-Saez AJ, Malovrh P, Bacia K, Anderluh G et al (2008) Equinatoxin II permeabilizing activity depends on the presence of sphingomyelin and lipid phase coexistence. Biophys J 95:691–698
18. Steringer JP, Bleicken S, Andreas H, Zacherl S, Laussmann M et al (2012) Phosphatidylinositol 4,5-bisphosphate (PI(4,5)P2)-dependent oligomerization of fibroblast growth factor 2 (FGF2) triggers the formation of a lipidic

membrane pore implicated in unconventional secretion. J Biol Chem 287:27659–27669

19. Bergstrom CL, Beales PA, Lv Y, Vanderlick TK, Groves JT (2013) Cytochrome c causes pore formation in cardiolipin-containing membranes. Proc Natl Acad Sci 110(16):6269–6274
20. Bleicken S, Garcia-Saez AJ, Conte E, Bordignon E (2012) Dynamic interaction of cBid with detergents, liposomes and mitochondria. PLoS One 7:e35910
21. Bleicken S, Classen M, Padmavathi PV, Ishikawa T, Zeth K et al (2010) Molecular details of Bax activation, oligomerization, and membrane insertion. J Biol Chem 285:6636–6647
22. Hermann et al (2014) Bioinformatics. doi: 10.1093/bioinformatics/btu102
23. Dimitrov DS, Angelova MI (1986) Swelling and electroswelling of lipids theory and experiment. Stud Biophys 113:15–20
24. Dimitrov DS, Angelova MI (1988) Lipid swelling and liposome formation mediated by electric-fields. Bioelectrochem Bioenerg 19: 323–336

Chapter 17

Purification of Recombinant 2XMBP Tagged Human Proteins from Human Cells

Ryan Jensen

Abstract

The ability to purify an intact, functional protein or protein complex is an essential step in biochemical characterization studies. Challenges in purification arise when proteins are of low abundance or are unstable resulting in low yields or poor in vitro activity. In this protocol we describe a method to purify active, recombinant human proteins fused to a tandem MBP tag after expression in human 293T cells.

Key words Protein purification, Recombinant human proteins, MBP tag, 293T cells, Unstable proteins, Mammalian expression systems

1 Introduction

The ability to overexpress and purify large quantities of recombinant human proteins for both basic research and therapeutic use is of paramount importance to biomedical researchers and the biotechnology/pharmaceutical industry [1]. The choice of expression systems in which to produce recombinant proteins ranges from bacteria to human cells and depends upon the size and complexity of the protein to be purified [2]. The purification of endogenous human proteins requires a large amount of starting material and normally requires multiple chromatography steps to acquire a homogenous protein with the required specific activity. In contrast, the development of numerous affinity tags combined with recombinant overexpression of a protein can yield sufficient quantities of pure protein in a simple one or two step purification scheme [3]. As long as the purification strategy yields enough active protein to allow for functional biochemical characterization, the researcher can avoid complex and more expensive methods that may or may not lead to success. The procedure we describe here requires only modest experience in mammalian cell culture, commonly acquired reagents and by using gravity-flow columns

Narendra Wajapeyee (ed.), *Cancer Genomics and Proteomics: Methods and Protocols*, Methods in Molecular Biology, vol. 1176, DOI 10.1007/978-1-4939-0992-6_17,

and a syringe pump does not necessitate the use of an expensive automated FPLC purification system. We utilized a protocol that facilitated the purification of the full-length human BRCA2 (breast cancer susceptibility gene 2) gene fused at the N-terminus with a tandem repeat of the MBP (maltose binding protein) tag [4]. Because 2XMBP is a large affinity tag (~84 kDa) relative to the size of most human proteins, we recommend incorporating a protease cleavage site to allow for removal in case the tag interferes with protein activity [5].

2 Materials

All buffers are prepared using Nanopure (Millipore or Barnstead) water, analytical grade reagents, and are 0.2 μM filtered prior to use (unless otherwise indicated). All purification steps are performed at 4 °C and access to a cold room is highly recommended. 1 mM DTT is added to buffers directly before use (*see* **Note 1**).

2.1 Mammalian Cell Transfection Reagents

1. Twenty 15-cm tissue culture plates (Corning, 430599).
2. Four 500 mL bottles of DMEM media (Life Technologies, 11965118) with 10 % FBS (Life Technologies, 26140079, heat inactivation is not recommended).
3. 293T cells (ATCC, CRL-11268).
4. 100 mM Chloroquine (Sigma, C6628): 0.52 g in 10 mL water. Filter and store in dark at −20 °C.
5. 250 mM $CaCl_2$ (Sigma, C5670): 8.32 g $CaCl_2$ in 300 mL Nanopure water.
6. 2× HBS (HEPES buffered saline): 2.4 g NaCl, 0.064 g Na_2HPO_4, 3.6 g HEPES, bring volume close to 300 mL with Nanopure water, pH to exactly 7.05 (*see* **Note 2**), bring up to 300 mL with water.

2.2 Cellular Extraction and Chromatography Reagents

1. Amylose resin (New England Biolabs, E8021S)
2. Glutathione Sepharose 4B (GE, 17-0756-01)
3. Four disposable plastic gravity-flow columns, 5 mL (Pierce, 29922)
4. HiTrap Q or SP (GE)
5. Syringe pump (Braintree Scientific, BS-8000 US)
6. Buffer H: 50 mM HEPES (pH 7.5), 250 mM NaCl, 5 mM EDTA, 1 % Igepal CA-630 (NP-40), protease inhibitor cocktail (Roche), and 1 mM DTT.
7. Buffer W (Wash): 50 mM HEPES (pH 7.5), 250 mM NaCl, 0.5 mM EDTA.

8. Buffer LS (Low Salt): 50 mM HEPES (pH 8.2), 250 mM NaCl, 0.5 mM EDTA, 10 % glycerol, and 1 mM DTT.
9. Buffer HS (High Salt): 50 mM HEPES (pH 8.2), 1 M NaCl, 0.5 mM EDTA, 10 % glycerol, and 1 mM DTT.
10. Maltose elution buffer: 50 mM HEPES (pH 8.2), 250 mM NaCl, 0.5 mM EDTA, 10 % glycerol, 1 mM DTT, and 10 mM maltose (Sigma, M5895, d-(+)-maltose monohydrate).
11. HiTrapQ elution buffer: 50 mM HEPES (pH 8.2), 450 mM NaCl, 0.5 mM EDTA, 10 % glycerol, and 1 mM DTT.

3 Methods (Table 1) (Also *See* Fig. 1)

3.1 Molecular Cloning of cDNA In-Frame with 2XMBP Tag

Prior to transfection, the human cDNA to be purified is cloned into the mammalian expression vector, phCMV1 (Genlantis, P003100). We have found that an in-frame fusion with two tandem repeats of the MBP tag (2XMBP) located at the N-terminus of the protein facilitates expression and solubility of most human cDNAs. Fusion at the C-terminus may also work and is subject to the discretion of the investigator.

1. We generated the 2XMBP tag in phCMV1 by PCR using the pMAL plasmid (malE, New England Biolabs) as a template [6]. The first MBP ORF contains an ideal Kozak consensus sequence including an extra glycine residue after the start methionine (GCCACC**ATG**GGG) to ensure translation initiation from the first MBP sequence [7]. The start ATG was deleted from both the second MBP tag and the cDNA of the protein to be purified.
2. A PreScission Protease (LEVLFQ/GP) cleavage site was engineered directly upstream of the cDNA to be purified [8]. The PreScission Protease recognition site located between

Table 1
Outline of procedure

Experimental timeline
Day 1 Seed 293T's into 20 × 15 cm plates
Day 2 (morning) Chloroquine pretreat 1 h, transfect cells
Day 3 (morning) Refeed cells
Day 3 (evening) Harvest cell extracts, amylose batch bind overnight
Day 4 (morning) Pour amylose into columns, wash, maltose elution, proceed to second column (e.g., HiTrapQ or SP), wash, elute, analyze fractions by SDS-PAGE, concentrate, aliquot, and freeze fractions at −80 °C

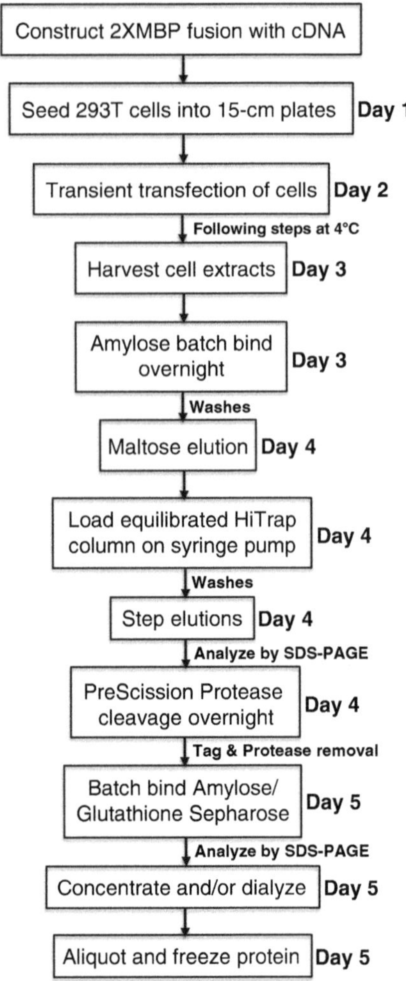

Fig. 1 Flowchart of 2XMBP fusion protein purification

the second MBP tag and the protein of interest will allow for subsequent cleavage of the entire 2XMBP tag if necessary.

3. If the cDNA is generated by PCR, the entire construct should be verified by DNA sequencing. Pay particular attention to the junction between the second MBP tag and the start of the cDNA.

4. The DNA for transfection can be purified using any standard Qiagen Maxi- or Mega-prep kit and should be stored in EB (Elution Buffer, 10 mM Tris–Cl pH 8.5) to enhance solubility and stability of the plasmid DNA.

3.2 Mammalian Cellular Transfection

Calcium phosphate-mediated transfection can be used to transfect 293T cells [9]. If cost is not an issue, commercially available reagents such as TurboFect (ThermoScientific) or FuGene (Roche) provide a simple and effective alternative. We have found that calcium phosphate-mediated transfection is equivalent to any commercially available cationic lipid based reagent in 293T cells.

1. Seed twenty 15-cm plates (~3×10^7 cells/dish) with 293T cells (DMEM + 10 % FBS media)
2. 24 h later, the cells should appear 60–70 % confluent. Treat the cells with 25 μM chloroquine for 1 h (138 μL 100 mM chloroquine stock in 550 mL DMEM + 10 % FBS). Aspirate media off cells and add media with chloroquine.
3. Use four 15-mL falcon tubes. In each tube, pipet 6.25 mL 250 mM $CaCl_2$ and then add 250 μg DNA. Mix. This is a 5× cocktail, for five 15-cm plates of 293T cells. Add drop-wise 6.25 mL of 2× HBS, bubble with pipet briefly, then invert to mix. Incubate for 5 min (*see* **Note 3**).
4. Add drop-wise gently to cell media (2.5 mL/15-cm plate). Swirl media in plate.
5. 24 h post-transfection, refeed the cells with fresh media.

3.3 Harvesting Cell Extracts and Purification

1. At 32–36 h post-transfection (the cells should be 70–90 % confluent at this point) bring plates into cold room and let equilibrate to 4 °C for about 15 min (*see* **Note 4**).
2. At this point, *all steps are performed at* 4 °C. Wash cell monolayers on plates twice with PBS, remove all PBS (*see* **Note 5**), and pipet 10 mL lysis buffer/15 cm plate. Place on rocker for 10 min. Scrape cell extracts using a rubber policeman into 50 mL tubes.
3. Place on lab rotator for another 10–20 min (*see* **Note 6**).
4. Transfer extracts to Oakridge tubes and centrifuge 10,000 × *g* for 15 min at 4 °C (e.g., Sorval SS-34). Make sure all tubes are prechilled to 4 °C.
5. After centrifugation is complete, carefully transfer supernatant to four prechilled 50 mL tubes (*see* **Note 7**).
6. *Batch binding:* Add 2 mL pre-washed, equilibrated amylose resin (*see* **Note 8**) per 50 mL cell extract.
7. Place on rotator overnight (*see* **Note 9**).

3.4 Assembly of Gravity-Flow Columns and Elution

1. The next morning: centrifuge the 50 mL tubes containing cellular extract and amylose resin for 5 min, 2,000 × *g* at 4 °C (*see* **Note 10**).
2. Remove unbound supernatant (*see* **Note 11**). Wash amylose resin with 10 mL buffer W (+1 mM DTT), and resuspend beads with pipet to wash.
3. Centrifuge again for 5 min at 2,000 × *g*, 4 °C.

4. Prepare four disposable plastic columns (Pierce) in cold room on ring stands, rinse with water several times (*see* **Note 12**), cap bottom, fill water to neck, push frit to bottom with plunger, decant water out.
5. Resuspend each of the four 50 mL tubes of amylose with 2 mL buffer W (+1 mM DTT) and pipet into column on top of frit, rinse tube with another 1 mL of wash buffer to get all amylose into column, let amylose settle and form a gel bed (*see* **Note 13**).
6. Place second frit in buffer on top of the column and use plunger to push down about 1 mm above amylose gel bed (*see* **Note 14**).
7. Wash all four columns with 40 mL of buffer W (+1 mM DTT) and repeat three times. Use the large column adaptors (Pierce, 29923) on top to accommodate this volume.
8. After each column has been washed with 120 mL of buffer W, elute with 4 mL of maltose elution buffer.
9. Elute 4 mL through the first column twice (let first 0.5 mL drip to waste beaker), then take that same 4 mL and use to elute the second column twice and do the same with the third and fourth columns (i.e., use the same 4 mL eluate for all 4 columns and run it through each column twice). To capture the residual 1 mL eluate trapped in the column after the second elution, place the column in a 15-mL tube and centrifuge $300 \times g$ for 3 min, repeat for all four columns (*see* **Note 15**).

3.5 Secondary Column Purification Using a Syringe Pump

If further purification is required to remove contaminants after maltose elution, several different 1 mL HiTrap chromatography columns are available (http://www.gelifesciences.com/webapp/wcs/stores/servlet/catalog/en/GELifeSciences/brands/hitrap/). Protein binding and salt-dependent elution is dependent upon multiple factors intrinsic to the protein of interest (e.g., net surface charge); choice of column and elution conditions is an empirical process. Here, we describe conditions specific to purification of the 2XMBP-BRCA2 protein. The following procedures are performed in the cold room.

1. Prepare a 1 mL HiTrapQ (GE) pre-packed column. Set syringe pump to dispense at 1 mL/min. Attach a 10 mL (BD) syringe to the column with the appropriate luer lock adaptor (normally provided with purchase of HiTrap columns). Wash the column with 10 mL of Nanopure water to remove column storage buffer (20 % ethanol). Wash with 10 mL Buffer LS, wash with 10 mL Buffer HS, and equilibrate the column with 10 mL Buffer LS (*see* **Note 16**).
2. Load the amylose eluate (approximately 4 mL), collect flow-through, and wash with 20 mL of Buffer LS.

3. Elute with 5 mL HiTrapQ elution buffer (*see* **Note 17**), collect 0.5 mL (or less) fractions in eppendorf tubes. 2XMBP-BRCA2 will come off very quickly after the first 1 mL has passed through the Q column (*see* **Note 18**).
4. Run all elutions and fractions on SDS-PAGE gel to determine where peak fractions are located and to check quality of the preparation (*see* **Note 19**).
5. Dialyze or desalt the protein as needed.
6. Pool and aliquot correct fractions. At this stage, the protein can be concentrated, if necessary, using Amicon (Millipore), Concentrator (Pierce), or similar device.
7. Determine the concentration of the protein by absorbance at 280 nm using a spectrophotometer or use a dye reagent compatible (e.g., Bradford assay) based method compared to a known standard.
8. "Snap-freeze" the final protein aliquots in liquid nitrogen and transfer to −80 °C freezer (*see* **Note 20**).

3.6 Proteolytic Cleavage of 2XMBP Tag

If desired, the 2XMBP tag can be removed by incubation with PreScission Protease (GE). After cleavage, a glycine–proline dipeptide will remain at the N-terminus of the protein.

1. Add 1 unit (~1 μL) of recombinant PreScission Protease for every 100 μg of target protein. Incubate at 4 °C for 5–16 h.
2. To remove the GST tagged PreScission Protease, free 2XMBP tag, and any uncleaved target proteins, incubate the extract with 0.2 mL of a glutathione sepharose 4B–amylose resin slurry mixture for 1–2 h (*see* **Note 21**).
3. Centrifuge at 2,000 × *g* and transfer supernatant to a fresh tube with a fresh mixture of glutathione sepharose 4B and amylose resin slurry. Repeat **step 2** above. This will facilitate removal of all tagged proteins (*see* **Note 21**). To prevent bead/resin carryover, apply the mixture to a disposable column with a frit that can be used to capture the beads and allow the cleaved protein to flow-through (*see* **Note 22**).
4. Analyze the cleavage process by SDS-PAGE. 2XMBP tag alone will migrate at approximately 84 kDa. GST-PreScission Protease will migrate at 46 kDa.

4 Notes

1. Buffers containing "old" (more than 24 h) DTT should be discarded as oxidized DTT absorbs at 280 nm and will interfere with spectrophotometric determination of protein concentration [10]. Beta-mercaptoethanol or TCEP can be used as a reducing agent in place of DTT.

2. In order for calcium phosphate-mediated transfection to be effective, the pH of the 2× HBS is critical. Upon first use and due to variability in the accuracy of different pH meters, it is advisable to generate three or four 2× HBS solutions with 0.05 pH unit steps around pH 7.05 and perform a pilot experiment to determine the optimal pH for transfection.
3. Calcium phosphate–DNA mixture should form a cloudy precipitate within a few minutes if working properly.
4. Allowing the plates to equilibrate to 4 °C will aid in preventing cells from sloughing off during the washes with PBS. 293T cells will detach easily if not handled carefully. We recommend tilting plates, begin pouring in a corner, then slowly let PBS cover cells followed by very gentle swirling, and then decant into waste container. Do not stack plates.
5. Make sure to remove all of the PBS wash from the cell monolayer, tilt plates by placing on lids at an angle to remove residual PBS wash.
6. Additional gentle agitation will allow the NP-40 to completely lyse the cells. Avoid frothing the extract from this point on.
7. Do not let any part of the pellet or precipitate decant into the new tubes (if it does by accident, remove it).
8. Before use, wash amylose resin several times to remove any traces of ethanol storage buffer.
9. Use rubber bands to secure tubes to rotator, parafilm around caps to prevent accidental leakage, make sure tubes are sufficiently filled with volume to prevent excess agitation and frothing, and ensure that tubes are balanced and will not fall off during the overnight batch binding.
10. It is not recommended to centrifuge amylose resin at speeds exceeding 2,000 × *g*.
11. Carefully decant supernatant as amylose resin "pellet" will be loose.
12. Ensure that new columns flow properly with water and that no manufacturing defects or particulates impede the flow.
13. Before the amylose settles, fill buffer past neck to top of column so that the top frit can be pushed down with the plunger without disturbing the gel bed. Let the gel bed form over a 15–20 min time period.
14. The second frit located above the gel bed will prevent the resin from drying out once the wash buffer has been depleted from above the column.
15. The amylose gel bed is desiccated after centrifugation so discard the column at this point.
16. HiTrap columns can be re-used several times if properly washed and stored after each use. Avoid introducing air bubbles into the

column by removing the syringe after each change of buffer, fill syringe completely, push buffer to end of syringe and tap to remove any remaining bubbles.

17. Mix 2 mL of buffer HS with 8 mL of buffer LS to make 450 mM NaCl final concentration.
18. It is advisable to collect small fractions (to keep the protein concentrated and partition contaminants).
19. SYPRO Orange (Sigma, S5692) stain can be used to visualize bands within 1 h, alternatively, Bio-Rad StainFree gels can be used to visualize bands immediately after electrophoresis using a UV transilluminator.
20. If additional glycerol (e.g., up to 50 %) is needed to stabilize the protein, add it to the storage buffer before freezing. Depending on the concentration of the protein and the specific assay to be utilized, now is the time to determine appropriate volumes to freeze such that the protein is not subjected to multiple freeze–thaw cycles.
21. The final amount of glutathione–amylose resin mixture needed and incubation time required to bind and remove tagged proteins will depend on the concentration of the target proteins and the binding capacity of the resins.
22. Pierce centrifuge columns (89868) are useful for capturing beads and collecting flow-through after a single centrifugation.

References

1. Goeddel DV (1990) Systems for heterologous gene expression. Meth Enzymol 185:3–7
2. Brondyk WH (2009) Selecting an appropriate method for expressing a recombinant protein. Meth Enzymol 463:131–147
3. Terpe K (2003) Overview of tag protein fusions: from molecular and biochemical fundamentals to commercial systems. Appl Microbiol Biotechnol 60(5):523–533
4. Jensen RB, Carreira A, Kowalczykowski SC (2010) Purified human BRCA2 stimulates RAD51-mediated recombination. Nature 467(7316):678–683
5. Charlton A, Zachariou M (2011) Tag removal by site-specific cleavage of recombinant fusion proteins. Methods Mol Biol 681: 349–367
6. di Guan C, Li P, Riggs PD, Inouye H (1988) Vectors that facilitate the expression and purification of foreign peptides in Escherichia coli by fusion to maltose-binding protein. Gene 67(1): 21–30
7. Kozak M (1987) An analysis of 5′-noncoding sequences from 699 vertebrate messenger RNAs. Nucleic Acids Res 15(20):8125–8148
8. Leong LE (1999) The use of recombinant fusion proteases in the affinity purification of recombinant proteins. Mol Biotechnol 12(3):269–274
9. Sambrook J, Russell DW (2006) Calcium-phosphate-mediated transfection of eukaryotic cells with plasmid DNAs. CSH Protoc 2006(1)
10. Cleland WW (1964) Dithiothreitol, a new protective reagent for SH groups. Biochemistry 3: 480–482

Chapter 18

Computational Analysis in Cancer Exome Sequencing

Perry Evans, Yong Kong, and Michael Krauthammer

Abstract

Exome sequencing in cancer is a powerful tool for identifying mutational events across the coding region of human genes. Here, we describe computational methods that use exome sequencing reads from cancer samples to identify somatic single nucleotide variants (SNVs), copy number alterations, and short insertions and deletions (InDels). We further describe analytical methods to generate lists of driver genes with more mutational events than expected by chance.

Key words Cancer, Single nucleotide variant, Copy number variation, Exome sequencing, Gene burden, InDels

1 Introduction

Exome sequencing of paired normal and tumor samples allows for the identification of somatic changes in a patient, including the identification of single nucleotide variants (SNVs), copy number changes, and short InDels (insertions and deletions). The falling cost of exome sequencing allows for the aggregation of such mutations across the coding regions of human genes, yielding a more complete picture of the mutational landscape of cancer. Here, we outline methods for utilizing such data to identify genes that might be important for cancer. We first present methods for preprocessing and alignment of sequencing reads, followed by methods that are useful for the identification of genes with more mutational events than expected by chance. The underlying methods need to address the many genomic biases that affect the number of observed mutations, including chromatin state, or the replication time of the genomic loci with the mutations. Our analytical methods use indirect evidence to estimate the magnitude of these biases, mainly in the form of neutral mutational events (i.e., synonymous mutations).

Narendra Wajapeyee (ed.), *Cancer Genomics and Proteomics: Methods and Protocols*, Methods in Molecular Biology, vol. 1176, DOI 10.1007/978-1-4939-0992-6_18, © Springer Science+Business Media New York 2014

2 Materials

For this analysis, we utilized existing software and developed our own scripts as necessary. All software used is freely available. Our sequencing data consisted of fastq files generated by Illumina GA2 and HiSeq machines, after exome capture using the NimbeGen SeqCap EZ Exome Library kit.

2.1 Data Files

1. Fastq files from sequencing runs for all normal and tumor samples.
2. A file indicating what genomic bases are covered in the exome capture region.

2.2 Software

1. Btrim—software to trim adapters and low quality regions in reads, obtained at http://graphics.med.yale.edu/trim/ [1].
2. BWA—Burrows-Wheeler Aligner, obtained at http://bio-bwa.sourceforge.net/ [2].
3. SAMtools—Sequence Alignment/Map tools, obtained at http://samtools.sourceforge.net/ [3].
4. Variant Effect Predictor—tool for predicting the functional consequences of known and unknown variants, downloaded from http://useast.ensembl.org/info/docs/variation/vep/index.html.
5. CONTRA—Copy Number Targeted Resequencing Analysis, obtained at http://sourceforge.net/apps/mediawiki/contra-cnv/index.php [4].
6. CMDS—correlation matrix diagonal segmentation, obtained at https://dsgweb.wustl.edu/qunyuan/software/cmds [5].
7. DNACopy—DNA copy number data analysis in R, obtained at http://www.bioconductor.org/packages/2.11/bioc/html/DNAcopy.html.
8. Circos—tool for visualizing data in a circular layout, obtained at http://circos.ca/ [6].
9. Integrated Genomics Viewer—tools for visualizing genomic tracks, downloaded from http://www.broadinstitute.org/igv/ [7].
10. LiftOver—tool to convert genome coordinates between assemblies, obtained at http://genome.ucsc.edu/cgi-bin/hgLiftOver.
11. R—language and environment for statistical computing and graphics, downloaded from http://cran.r-project.org/.

2.3 Databases

1. UCSC GRCh37/hg19 human genome sequence and RefGene.txt file (*see* **Note 1**).
2. dbSNP—NCBI database of short genetic variations, obtained at http://www.ncbi.nlm.nih.gov/SNP/ or http://genome.ucsc.edu/.
3. 1000 Genomes—catalog of human genetic variation, obtained at ftp://ftp-trace.ncbi.nih.gov/1000genomes/ftp/.

3 Methods

The methods outlined below were developed while analyzing 99 melanoma samples and their matched normal samples [8]. These methods are applicable to sequence data generated by any next-generation sequencing and capture platforms. Figure 1 shows an overview of the methods.

3.1 Alignment

1. For each exome fastq file, trim low quality regions from reads using Btrim with a window size of 10 and a quality cutoff of 25. If adapters exist, they should be trimmed by Btrim.
2. Align trimmed reads to GRCh37/hg19 using BWA with default parameters.
3. For each sample, find the average sequencing error rate as the fraction of bases from sequencing reads that do not align with the reference genome.

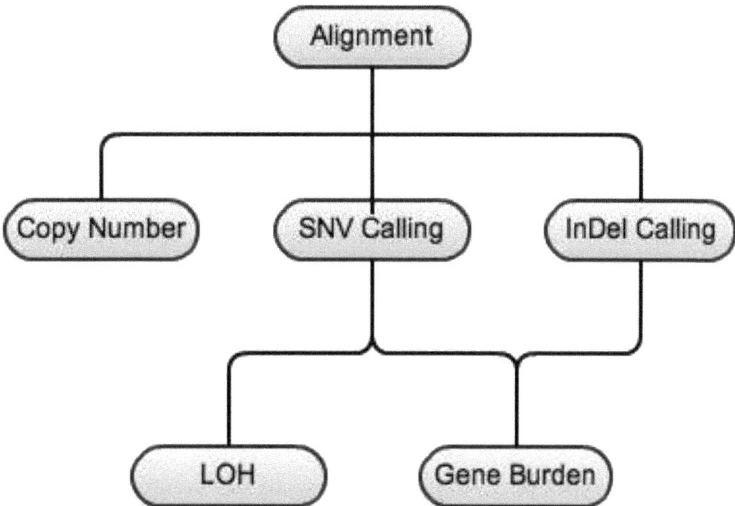

Fig. 1 Sequence analysis pipeline. For each sample, reads are aligned to the UCSC human genome. Following alignment, sequence data are used to produce copy number variations, single nucleotide variants (SNVs), and insertion/deletions (InDels). SNV data are used to find loss of heterozygosity regions in tumors, and determine genes with significant nonsynonymous SNV burden across samples

4. For each sample, determine the average coverage for each base in the capture region.
5. For each sample, find the percent of bases covered at least eight times across the capture region.
6. Use the three sequence quality measures above to discard samples where coverage is too low, or the error rate is too high (*see* **Note 2**).

3.2 SNV Calling

1. Use SAMtools to remove PCR duplicates and call single nucleotide variants (SNVs).
2. Annotate SNV protein effects (synonymous, nonsynonymous, UTR, intron) based the annotation files from UCSC. RefGene.txt from UCSC contains the exon, intron, as well as transcript start and end positions. Using this information, the phase of each nucleotide with each codon can be determined. Use the Variant Effect Predictor on intronic SNVs to identify splice-site variants.
3. Check synonymous and nonsynonymous annotations for adjacent pairs of SNVs affecting the same codon. If present, sequencing reads are scanned for the occurrence of both SNVs on the same read, and the amino acid change is predicted based on the simultaneous mutations.
4. Filter SNVs as follows. Keep SNVs that meet these criteria: at least a 13 % mutant allele frequency, at least a SAMtools mapping score of 40, minimum coverage of four mutant and eight total reads, at least one forward and one backward read, and a uniform mapping of reads with the mutant allele.
5. Produce a list of novel SNVs by filtering out SNVs present in dbSNP and 1000 Genomes (*see* **Note 3**).
6. An SNV is somatic if the following criteria hold: one or less mutant read in the normal samples, and a sufficient variant to total read ratio in tumor and normal samples as assessed by Fisher's exact test (p-value threshold of 0.001).
7. There will be cases where a tumor has a clear SNV, but the normal coverage at this position is too low to determine if the SNV is somatic or present in both tumor and normal. For these cases, the SNV will be called somatic if it is somatic in another sample.

3.3 Somatic SNV Call Precision and Sensitivity

1. Randomly choose a few hundred SNVs called in tumors and validate these with Sanger sequencing (*see* **Note 4**). Call the set of validated SNVs V.
2. Calculate the precision of SNV calling as the percent of called SNVs that were validated by Sanger sequencing.

3. Find the subset of *V* that is called as somatic by the computational pipeline. Call this set *SV*. Use Sanger sequencing to test for the absence of all SNVs in *SV* in the matched normal samples. This test indicates how many SNVs are truly somatic, *ST*. Calculate the precision of somatic SNV calls as *ST*/*SV*, the percent of SNVs called as somatic that were validated to be somatic by Sanger sequencing.
4. In each normal sample, identify variants present in dbSNP. Calculate the sensitivity of SNV calls as the percent of normal SNPs called in the corresponding matched tumor samples (*see* **Note 5**).
5. Take all SNVs from *V*, and test normal samples with Sanger sequencing to see if the SNVs are also present in the matched normal samples. Retain SNVs that are truly somatic, and calculate the sensitivity of the somatic calling pipeline as the percent of truly somatic SNVs detected computationally.

3.4 InDel Calling

1. Use SAMtools to call InDels.
2. Filter for InDels with a mutant allele frequency greater than 0.1, a minimal SNP score of 250 in at least one tumor sample, and an absence from dbSNP.
3. Find somatic InDels by excluding all InDels present in normal samples. When an InDel is not annotated in a normal sample, require the normal sequence coverage to be at least 8 independent reads for the tumor InDel to pass the filter.
4. Use the Variant Effect Predictor on all InDels to determine their effects. Retain InDels that cause frame-shifts or affect stop codons.

3.5 Somatic Copy Number Analysis

1. Use CONTRA with the multimapped read exclusion flag to determine copy number events for each matched melanoma sample.
2. The sequence coverage log fold change between tumor and normal produced by CONTRA can be visualized with the Integrated Genomics Viewer.
3. Count the number of samples for which at least one exon in a gene has a significant CONTRA call and fit a Poisson distribution to the resulting sample counts per gene. Retain genes that have copy number events in significantly more samples than expected.
4. Run CMDS across all samples. Limit genes with CONTRA copy number calls to those that are located in chromosomal bands with significant CMDS calls to produce a final list of enriched somatic copy number events.

3.6 Loss of Heterozygosity

1. Identify heterozygous positions in normal samples as SNVs with a mutant allele frequency ranging from 0.4 to 0.6 and a sequence depth of at least 10. For each normal heterozygous SNV, record the zygosity of the matched tumor SNV. Tumor SNV zygosity is determined using a binomial test to compare the observed mutant allele frequency to 0.5. Use DNACopy to perform circular binary segmentation on homozygous and heterozygous positions in tumor.
2. Filter each DNACopy call for region-wide LOH by requiring that a high percentage of tumor SNVs within the region be labeled at homozygous.
3. For a summary visual of LOH regions, use Circos to view each sample's LOH regions in a circular layout.
4. Use LOH regions to determine the admixture of normal and tumor cells within a tumor sample as follows. For all tumor SNVs within LOH regions, determine the average absolute difference between each SNV's mutant allele frequency and 0.5. Double this average to obtain a measure of tumor sample purity. Pure samples will have a value close to one. Samples with low purity can be discarded in further analyses.

3.7 Identification of Genes with Significant Somatic Nonsynonymous Mutation Burden

1. Construct a file of simulated mutations as follows. For each base in the exome capture region, mutate to all possible nucleotides, and record the amino acid encoded by the new codon.
2. If mutations in your data have a genomic context bias, divide the simulation file into different contexts to account for mutation biases (*see* **Note 6**).
3. For each gene in the exome capture region, and for each genomic context, use the simulation file to determine the ratio of nonsynonymous to synonymous changes for the gene if all possible mutations occurred according to the observed distribution of somatic mutations (*see* **Note 7**).
4. For each mutated gene and each genomic context, find the expected number of nonsynonymous mutations by multiplying the observed number of synonymous mutations by the ratio of nonsynonymous to synonymous calculated above. Obtain a *p*-value for each genomic context using a binomial test to compare the observed number of nonsynonymous mutations to the expected number given the length of the gene for this context (*see* **Note 8**).
5. For each gene, combine *p*-values across genomic contexts using Fisher's combined probability test to arrive at a final *p*-value for the gene.
6. *P*-values must be corrected for multiple testing. We used the Benjamini–Hochberg procedure, as implemented in R through the BH function provided in the Mutoss package.

We entered *p*-values for every gene in the capture region. Genes with no nonsynonymous mutations were given a *p*-value of 1.

7. Some tumors may have many nonsynonymous mutations in the same gene, resulting in a high mutation burden ranking that is only attributable to a few samples. In this case, the gene will not be a relevant driver for the whole sample set. You can put an additional filter on your gene list to require that a gene is mutated in a minimum number of samples to avoid this problem.
8. To ensure that your gene list is robust, you can remove the top 5 % of 10 % of tumors by nonsynonymous SNV count and run the mutation burden analysis on this slightly smaller set of tumors. You can report genes that are ranked highly in both burden runs.
9. To further narrow down this final list of significant genes, compare the gene list with LOH regions and genes known to be expressed in tumors. Genes that have a significant mutation burden, are expressed in tumors, and are often in LOH regions are the good candidates for cancer drivers.
10. If your tumor samples can be divided into different classes, you might want to run this mutation burden for each sample class, provided there are enough somatic mutations. As an example, for melanoma tumors, we split our tumors based on sun exposure. We only ran the burden analysis on sun-exposed tumors because sun-shielded tumors had too few mutations to adequately fill the genomic contexts for all genes (*see* **Note 9**).

3.8 Identification of Genes with Significant Mutations that Abrogate Protein Function

1. For a special case of mutation burden, gather the nonsense mutations, splice-site variants, frame-shift InDels, and InDels that affect stop codons for all samples. For each gene, calculate the sum of these three mutation types over all samples (*see* **Note 10**).
2. For each gene, use a binomial test to find the probability of observing the deleterious mutation count given the gene length and the exome-wide probability of a deleterious mutation occurring. For gene length, use the length of the gene in the capture region.
3. Use the Benjamini–Hochberg procedure to correct for multiple testing, and apply any of the additional filters from the nonsynonymous gene burden test above as needed.

4 Notes

1. All genomic coordinates for this work reference hg19. When a dataset, like 1000 Genomes, was not represented in hg19 coordinates, use liftOver to convert coordinates to hg19.

2. Melanomas showed an average sequence error rate around 0.24 %, an average coverage of 65 reads (minimum 30 and maximum 93), and at least 90 % of the capture region bases were covered by at least 8 reads.
3. dbSNP contains some rare important cancer mutations, like BRAF V600E at chr7:140453136 and chr7:140453137. dbSNPs like these are marked as having an unknown origin in dbSNP. Unknown variants should not be considered in the novel filtering process.
4. For melanoma, we validated 266 SNVs with Sanger sequencing.
5. We assume that variant normal bases that correspond to SNPs will rarely be mutated in tumor samples.
6. As an example, melanoma has three relevant genomic contexts: dipyrimidine cytosines, non-dipyrimidine cytosines, and thymidines (and their complements).
7. To find the observed distribution of mutations you can count either synonymous mutations or both nonsynonymous and synonymous mutations. Using only the synonymous mutations is preferred because nonsynonymous mutation might be affected by selection. If nonsynonymous mutations are used to determine the observed mutation distribution, do not use nonsynonymous mutations from the top most frequently mutated genes according to nonsynonymous mutation count.
8. If a gene has no synonymous mutations for a context, use the genome-wide frequency of synonymous mutations for this context with the gene's length and nonsynonymous to synonymous ratio to arrive at an expected number of nonsynonymous mutations.
9. For sun-exposed melanomas, a median of 171 nonsynonymous mutations across 61 tumors was sufficient to run the mutation burden analysis, but a median of 9 nonsynonymous mutations across 26 sun-shielded tumors resulted in too few mutations. To produce a candidate gene list of cancer drivers for tumors with few mutations, you can simply find genes with recurrent nonsynonymous mutations from different samples that affect the same codon.
10. These mutations have a higher probability of affecting protein function than mutations that change one amino acid to another, and should be analyzed separately.

Acknowledgements

This work was supported by the Yale SPORE in Skin Cancer funded by the National Cancer Institute grant number 1 P50 CA121974 (principal investigator, Ruth Halaban), the Melanoma

Research Alliance (a Team award to Ruth Halaban and M.K.), The National Library of Medicine Training grant 5T15LM007056 (P.E.), Yale Comprehensive Cancer Center (M.K.), and Gilead Sciences, Inc. (M.K.).

References

1. Kong Y (2011) Btrim: a fast, lightweight adapter and quality trimming program for next-generation sequencing technologies. Genomics 98(2):152–153
2. Li H, Durbin R (2009) Fast and accurate short read alignment with Burrows–Wheeler transform. Bioinformatics 25(14):1754–1760
3. Li H et al (2009) The sequence alignment/map format and SAMtools. Bioinformatics 25(16):2078–2079
4. Li J et al (2012) CONTRA: copy number analysis for targeted resequencing. Bioinformatics 28(10):1307–1313
5. Zhang Q et al (2010) CMDS: a population-based method for identifying recurrent DNA copy number aberrations in cancer from high-resolution data. Bioinformatics 26(4):464–469
6. Krzywinski M et al (2009) Circos: an information aesthetic for comparative genomics. Genome Res 19(9):1639–1645
7. Thorvaldsdottir H, Robinson JT, Mesirov JP (2012) Integrative Genomics Viewer (IGV): high-performance genomics data visualization and exploration. Brief Bioinform 14(2):178–192
8. Krauthammer M et al (2012) Exome sequencing identifies recurrent somatic RAC1 mutations in melanoma. Nat Genet 44(9):1006–1014

Chapter 19

Matrix Factorization Methods for Integrative Cancer Genomics

Shihua Zhang and Xianghong Jasmine Zhou

Abstract

With the rapid development of high-throughput sequencing technologies, many groups are generating multi-platform genomic profiles (e.g., DNA methylation and gene expression) for their biological samples. This activity has generated a huge number of so-called "multidimensional genomic datasets," providing unique opportunities and challenges to study coordination among different regulatory levels and discover underlying combinatorial patterns of cellular systems. We summarize a matrix factorization framework to address the challenge of integrating multiple genomic datasets, as well as a semi-supervised variant of the method that can incorporate prior knowledge. The basic idea is to project the different kinds of genomic data onto a common coordinate system, wherein genetic variables that are strongly correlated in a subset of samples form a multidimensional module. In the context of cancer biology, such modules reveal perturbed pathways and clinically distinct patient subgroups that would have been overlooked with only a single type of data. In summary, the matrix factorization framework can uncover associations between distinct layers of cellular activity and explain their biological implications in multidimensional data.

Key words Bioinformatics, Cancer genomics, Machine learning, Non-negative matrix factorization, Multidimensional genomics data

1 Introduction

Cellular systems have multiple levels of organization, such as epigenetic status, transcriptions, translations, transportation, and metabolic reactions. All these levels can interact and influence each other. Discovering their precise relationships and means of coordination is essential if we are to understand the function and organization of cellular systems. However, so far most research has focused on profiling only one level of organization at a time, due to a lack of appropriate data resources. With the fast development of high-throughput technologies, it is becoming possible to characterize large samples of biological systems on multiple levels simultaneously. For example, The Cancer Genome Atlas (TCGA) project aims to generate multidimensional genomic profiles for a

Narendra Wajapeyee (ed.), *Cancer Genomics and Proteomics: Methods and Protocols*, Methods in Molecular Biology, vol. 1176, DOI 10.1007/978-1-4939-0992-6_19, © Springer Science+Business Media New York 2014

set of patient samples including more than 20 types of tumors [1], and the NCI60 project has profiled 60 human cancer cell lines with multidimensional data [2]. As sequencing costs fall, multidimensional genomics characterization is becoming standard practice.

The format and structure of multidimensional genomics data pose new challenges for computational biologists. The data cannot be simply aggregated for further analysis, due to the diversity of types, scales, units, and systematic effects. Most relevant efforts have focused on integrating a single type of data obtained from different labs or cells, or on analyzing two-dimensional genomic datasets. Examples of the latter approach include traditional eQTL methods, which identify regulatory SNPs by jointly analyzing single-nucleotide polymorphism and gene expression data [3]; multivariate regression between gene expression and transcription factor binding data [4]; and the Ping-Pong algorithm, which integrates gene expression and drug response data [5].

Very few methods have been developed that can analyze multidimensional genomic datasets directly, regardless of the specific data types [6]. For example, a multivariate regression model has been employed to analyze multidimensional genomic data in a supervised manner. A biclustering method named cMonkey has been developed to analyze gene expression matrices from two different species. Finally, several multiple kernel learning methods have been designed for integrating heterogeneous genomic data, but these are only applicable to one or two matrices at a time.

The general goal of clustering methods is to identify a group of related rows or columns in a data matrix. The related task "biclustering" (or co-clustering) detects the simultaneous clustering of rows and columns in a data matrix [7]. Kutalik et al. [5] extended the traditional modular analysis approach to identify drug–gene co-modules. Shen et al. [8] proposed a joint clustering model for multiple genomic datasets, but it was designed for sample clustering and subtype discovery, and cannot identify modules comprising correlated variables.

We believe that simultaneous data integration will play a key role in exploring the coordination among multiple types of genomic information. Recently, we have proposed a powerful non-negative matrix factorization framework [9] and a semi-supervised variant [10] to identify correlative relationships in multidimensional genomics and network datasets (Fig. 1). These methods break down the data into building blocks that exhibit similar patterns across certain rows and columns of the matrices (Fig. 1). Within each building block, all or a subset of the samples exhibit correlated profiles across multiple matrices (Fig. 1). These building blocks are termed multidimensional modules (md-modules).

Matrix factorization methods can greatly reduce the complexity of a dataset, by providing a global atlas of the most significant associations among different types of genomic variables. The md-modules can be used to reveal significantly disrupted pathways and stratify patients into clinically distinct groups.

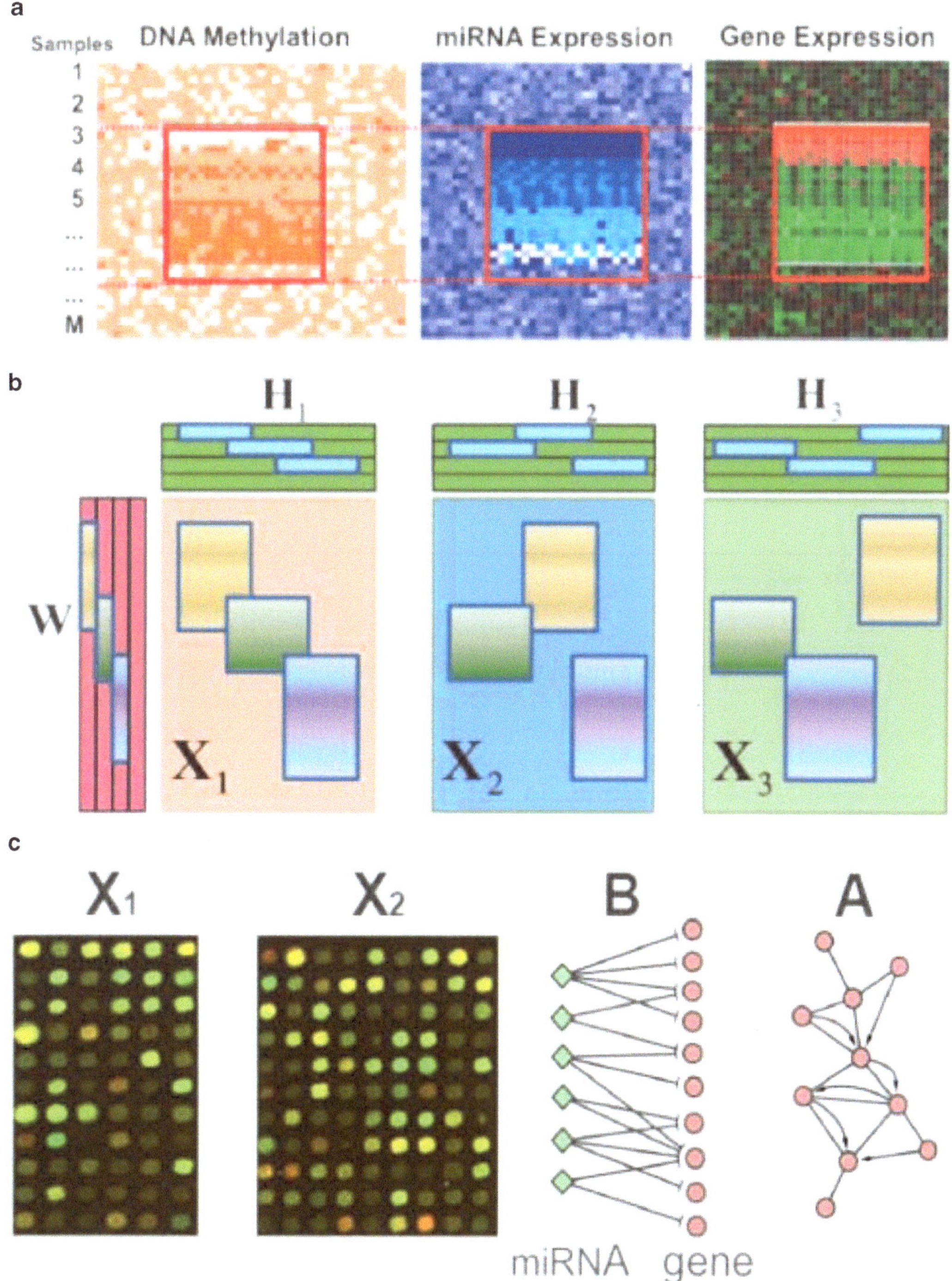

Fig. 1 (**a**) Multidimensional data and a multidimensional module (md-module). The matrices are TCGA profiles of 385 ovarian cancer samples. (**b**) Illustration of joint NMF factorization and three identified md-modules. (**c**) Illustration of data for the semi-supervised joint NMF factorization method

2 Materials

We downloaded our data from the TCGA Data Portal on April 27, 2009. We used three types of data: gene expression (Agilent G4502A), DNA methylation (Illumina 27K), and miRNA expression (Agilent H-miRNA_8x15K v2). A total of 385 samples have all three datasets. We normalized the columns of the three expression matrices respectively, and then scaled the matrices so that the sum of squares of each matrix is the same.

We employed the method suggested by Kim and Tidor [11] to make the input data non-negative. We doubled the columns of the microRNA and gene expression matrices, so that each profile element (gene, miRNA) is represented by two columns in the respective matrix. If the original value of the variable was positive, then it is stored in the first column; otherwise, its absolute value is stored in the second column. The remaining elements are filled with zeros.

3 Methods

Nonnegative matrix factorization (NMF) has become a very popular machine learning method in many fields, including bioinformatics [11–13]. NMF approximately factors a nonnegative matrix $X_{M \times N}$ into two nonnegative matrices, $X \approx WH$, where W is an $M \times K$ matrix of basis vectors and H is a $K \times N$ matrix of coefficient vectors [14, 15]. Typically, an appropriate solution is found by iteratively updating W and H in order to minimize an error function. Because each element of W and H must be ≥ 0, a key feature of NMF is its ability to identify non-subtractive patterns that explain the data as a linear combination of basis vectors. The K basis vectors in W can be regarded as "building blocks" of the data, while the K coefficient vectors describe how strongly each "building block" is present. The key expectation of NMF is that for any good solution (i.e., one where the difference between X and WH is small), each basis vector in W reproduces some statistically significant correlation between a subset of columns in X. Conversely, if such correlations exist, then NMF can find basis vectors that represent them and the factorization reproduces most of the information in the original dataset in fewer dimensions. This principle is illustrated in Fig. 1b. Note that the correlated elements need not be adjacent to each other in the matrix.

NMF has many bioinformatics applications, such as the clustering analysis of high-dimensional gene expression data [11–13]. Several variants of NMF have been proposed to enforce additional properties on W and H by incorporating discriminative constraints, locality-preserving or network-regularized constraints, sparsity constraints, and so on. Recently, an NMF-type method has been proposed to analyze gene expression and transcription factor-binding data [16].

We adapted the powerful NMF method for the discovery of modules in heterogeneous, multidimensional cancer genomic data profiled on the same samples [9]. We first introduced the idea using a three-dimensional dataset, but it is also applicable to higher-dimensional datasets. We then developed a semi-supervised framework for combining miRNA/gene expression profiles and network data to extract miRNA–gene regulatory co-modules [10].

3.1 The NMF Problem

Given a data set consisting of N measurements of M nonnegative scalar variables, the set of M-dimensional measurement vectors $x_{.j}$ ($j=1,\ldots,N$) forms the data matrix $X_{M\times N}$. For each column $x_{.j}$, a linear, nonnegative approximation of the data is given by $x_{.j}\approx \sum_{k=1}^{K} w_{.k}h_{kj}= Wh_{.j}$, or $X\approx WH$. where W is an $M\times K$ matrix containing the basis vectors $w_{.k}$ as its columns, and H is an $K\times N$ matrix containing the coefficient vectors $h_{.j}$ that approximate the measurement vectors $x_{.j}$. Note that each measurement vector is decomposed into the same set of basis vectors. The K basis vectors $w_{.k}$ can be thought of as the "building blocks" of the data, while the K-dimensional coefficient vector $h_{.j}$ describes how strongly each building block is present in the measurement vector $x_{.j}$.

Given a nonnegative data matrix X, the optimal choice of W and H is that which minimizes the reconstruction error between X and WH. Although several error functions have been proposed [15], the most widely used is the squared Euclidean error function:

$$F(W,H)=\|X-WH\|_F^2.$$

The W and H that minimize F is called the nonnegative matrix factorization of X. The best choice for the number of basis vectors K is often problem-dependent. In most cases, K is chosen such that $K<\min(M,N)$, so that WH represents a compressed form of the data in X. By not allowing negative entries in W and H, NMF describes a non-subtractive relationship between the basis vectors and the underlying dataset [15]. The following section describes how NMF can be extended to multiple datasets.

3.2 The Joint NMF Framework for Integrative Analysis

Let X_1, X_2, X_3 be $M\times N_1$, $M\times N_2$, and $M\times N_3$ matrices representing three types of genomic profiling of the same samples. For example, they can represent the methylation profiles of N_1 DNA markers and the expressions of N_2 genes and N_3 miRNAs, all obtained for the same M samples. To extract md-modules, we decompose the three data matrices into a common $M\times K$ basis matrix W. Thus, we obtain three different coefficient matrices H_I ($I=1,2,3$) such that $X_I\approx WH_I$, with the nonnegativity constraints $W\geq 0$ and $H_I\geq 0$. Each H_I is a matrix of size $K\times N_I$, and each row of H_I represents a coefficient vector. The Euclidean objective function is:

$$\min\sum_{I=1}^{3}\|X_I-WH_I\|_F^2.$$

3.3 The Semi-supervised Joint NMF Framework

To demonstrate the semi-supervised joint NMF framework, we apply it to identify miRNA–gene co-modules [10] (a co-module is a two-dimensional md-module in our terminology). The datasets are the non-negative miRNA and gene expression matrices X_1 and X_2, and the objective function is

$$\sum_{I=1}^{2} \| X_I - WH_I \|_F^2 .$$

The joint NMF solution is often not unique, and may be sensitive to noise in the biological data. Because both these limitations may confound the module discovery process, we guide the optimization process toward a reasonable biological solution by incorporating prior knowledge into the objective function. Specifically, we incorporate terms representing agreement with two types of network data: known gene–gene interactions and predicted miRNA–gene interactions (Fig. 1c). The essence of our semi-supervised learning framework is to define constraints such that any variables that are linked in these two datasets are more likely to be placed in the same co-module. In addition to improving the biological relevance of the results, such constraints can greatly facilitate the discovery of co-modules by narrowing down the large search space.

Let A denote the adjacency matrix of a gene interaction network, and B denote the adjacency matrix of a bipartite miRNA–gene network. We enforce "must-link" constraints by maximizing the following objective function:

$$O_1 = \sum_{ij} a_{ij} \left(h_i^2\right)^T h_j^2 = Tr\left(H_2 A H_2^T\right).$$

This term ensures that genes with known interactions have similar coefficient profiles. Similarly, the interactions between genes and miRNAs can be encoded by maximizing

$$O_2 = \sum_{ij} b_{ij} \left(h_i^1\right)^T h_j^2 = Tr\left(H_1 B H_2^T\right).$$

Our inputs are the miRNA and gene expression matrices X_1 and X_2 with dimensions $s \times m$ and $s \times n$ respectively, an $m \times n$ matrix B of predicted miRNA–target interactions, and an $n \times n$ matrix A representing the gene–gene interaction network. We combine the three objectives into a single optimization function:

$$F_1(W, H_1, H_2) = \sum_{I=1,2} \| X_I - WH_I \|_F^2 - \lambda_1 Tr\left(H_2 A H_2^T\right) - \lambda_2 Tr\left(H_1 B H_2^T\right)$$

The parameters λ_1 and λ_2 are weights for the must-link constraints defined in A and B. The first term favors modules with miRNA and gene expression profiles that are correlated in the common basis matrix W. The second term, $Tr(H_2AH_2^T)$, summarizes all the must-link constraints in the gene–gene network. The third term, $Tr(H_1BH_2^T)$, summarizes all the must-link constraints in the miRNA–gene network.

Although the NMF method often generates sparse representations of data, which aid in the discovery of part-based patterns [15], studies have shown that the specific representation is sensitive to the quality of the data and the choice of algorithm [10]. Several methods have been proposed to control the degree of sparseness in W and/or H [17]. In our semi-supervised joint NMF framework, we adopted a strategy suggested by Kim and Park [17] to make H_1 and H_2 sparse. The objective function is formulated as follows:

$$F_1(W, H_1, H_2) = \sum_{I=1,2} \|X_I - WH_I\|_F^2 - \lambda_1 Tr(H_2AH_2^T) - \lambda_2 Tr(H_1BH_2^T) + \gamma_1\|W\|_F^2 + \gamma_2\left(\sum_j \|h_j\|_1^2 + \sum_{j'} \|h_{j'}\|_1^2\right),$$

where h_j and h_j' are the jth and j'th columns of H_1 and H_2, respectively. The term $\gamma_1\|W\|_F^2$ limits the growth of W, while $\gamma_2\left(\sum_j \|h_j\|_1^2 + \sum_{j'} \|h_j'\|_1^2\right)$ encourages sparsity.

3.4 Algorithms

In the literature, several optimization algorithms have been developed for the NMF problem [15, 18]. The multiplicative algorithm designed by Lee and Seung [18] is simple to implement and performs well. Given a desired rank K, the method iteratively computes approximations of X_1, X_2, and X_3 to minimize the objective function. The algorithm starts by randomly initializing the matrices W and H_1, H_2, and H_3.

This is a local optimization procedure, so the algorithm converges to the first local minimum that it detects. We overcome this limitation by repeating the procedure multiple times (e.g., 50) with different realizations of the random initial matrices. The factorization with the smallest value of the objective function is used for further analysis. The solutions are reproducible in the sense that different runs of the repeated algorithm showed strong correlations. The time complexity of the joint NMF algorithm is $O(tK(M+N_1+N_2+N_3)^2)$, where t is the number of iterations. This is the same order as the original NMF algorithm. Therefore, given enough memory, it is applicable to large-scale matrices. If necessary, it is also possible to reduce the effective dimension of the input matrices by data-reduction techniques before applying the NMF algorithm.

The examples described above [9, 10] produce three coefficient matrices H_1, H_2, and H_3 for the joint NMF algorithm, or two matrices H_1 and H_2 for the semi-supervised version. Both algorithms can easily be extended to higher dimensionality. The coefficient matrices are then used to determine the membership functions of significant md-modules. In previous NMF applications, researchers have assigned a unique module to each gene (or other object) using the maximum element of each column in *H*. However, some genomic variables may not be active in any module, while others may be active in multiple modules. Therefore, for each row of *H* we calculate the *z*-scores of all elements and assign membership to all elements whose *z*-scores exceed a given threshold [9, 10]. In this way, the memberships of the modules (as represented in the basis vectors of *W*) can overlap. This principle is illustrated in Fig. 1b.

3.5 Statistical Significance of md-Modules

We expect that within an md-module, the profiles of different genomic variables are strongly correlated or anti-correlated. We suggest the following criterion to determine whether such relationships are statistically significant. First, calculate the "between-correlation" of two matrices with the same number of rows as the sum of the absolute values of Pearson's correlations between any two columns (one column from each matrix). Second, derive the statistical significance (*p*-value) of the between-correlation by calculating its distribution for 1,000 random matrix pairs obtained by permuting the elements of the original matrices.

3.6 Functional Analysis of Identified md-Modules

For each md-module, we can identify three gene sets of interest: (1) member genes from the GE dimension, (2) genes in the 20-kb region around member methylation markers in the DM dimension, and (3) genes targeted by member miRNAs in the ME dimension (based on miRNA targets listed in the Microcosm database). For each gene set, we performed an enrichment analyses in terms of GO biological processes and KEGG pathways. We can also perform enrichment analyses for cancer genes and protein interactions, and test the overlap between gene sets based on different variables.

3.7 Clinical Characterization

Recall that each column of the common basis matrix *W* contains coefficients representing distinct samples. We can characterize the level of association between each sample and the discovered md-modules or co-modules as follows. For each column of *W*, we divide the set of samples into two (or more) groups: those specific to a module and those not belonging to any module, based on a *z*-score threshold calculated for the column. Kaplan–Meier curves for the two groups were computed using *R*. Survival distributions between groups were computed via the log-rank test. Age differences between groups were compared by the Wilcoxon signed-rank test.

4 Results

The TCGA ovarian cancer dataset consisting of gene expression, DNA methylation, and miRNA expression profiles across 385 samples (patients) was used to demonstrate the discovery of multidimensional modules (md-modules). After parameter optimizations [9], the three large matrices were factored into $K=200$ building blocks, from which 200 (md-modules) were derived. This dimension reduction captures all the strongest correlations embedded in the original data. We calculated the average between-sample correlation between the reconstructed datasets (based on W and H_I) and the original datasets, obtaining values of 0.90, 0.92, and 0.91 in the methylation, miRNA, and gene expression dimensions respectively.

Each md-module comprises a set of genes, methylation markers, and miRNAs. Taken together, the 200 md-modules cover 2,985 genes, 2008 DNA methylation markers, and 270 miRNAs. The average module sizes in the gene, methylation marker, and miRNA dimensions are 239.6, 162.3, and 13.8, respectively.

4.1 Multidimensional Modules Reveal Cooperative Functional Effects

We first tested the functional homogeneity of member genes within each dimension. Among the 200 md-modules, 80 % had member genes that were functionally homogenous, and 62.7 % were homogenous with respect to genes that are directly adjacent to a member DNA methylation marker. Only 12.5 % of modules were homogenous with respect to genes targeted by member miRNAs. All three proportions are significantly higher than the frequency of enrichment obtained after randomization (5 %, 13.1 %, and 3.9 % for GE, DM, and ME respectively) (Fig. 2a). More interestingly, combining all three dimensions reveals more functional synergy. When the three sets of genes are combined, we find that 93 % of the md-modules are functionally homogenous, compared to only 7.9 % after randomization (Fig. 2). This result shows the power of integrative analysis in identifying genomic variables of different natures that are involved in the same functional pathways. The ability of the modules to capture multilevel synchronicity was also observed relative to perturbed KEGG pathways.

We expect an md-module to capture vertical associations, i.e., strong correlations between variables of different dimensions. Compared to randomly permuted modules, the Pearson's correlation coefficients between variables from two different dimensions (GE, DM, or ME) are significantly high. This result indicates that the probability of identifying these modules by chance is close to zero. The strong statistical correlations across different dimensions imply the coordinated activities of different genomic variables.

In 44 modules, genes from the GE and DM dimensions are separately enriched in protein–protein interactions (Fig. 2b) with p-values <0.05. Among these 44 modules, 18 are also enriched in protein–protein interactions bridging the GE and DM dimensions (p-value <0.05, calculated with the right-tailed Fisher's exact test). This finding highlights the different regulatory effects on closely connected molecules of the same pathway.

Finally, we hypothesized that some of the identified md-modules play a role in cancer. Indeed, 22 combined sets of genes (including

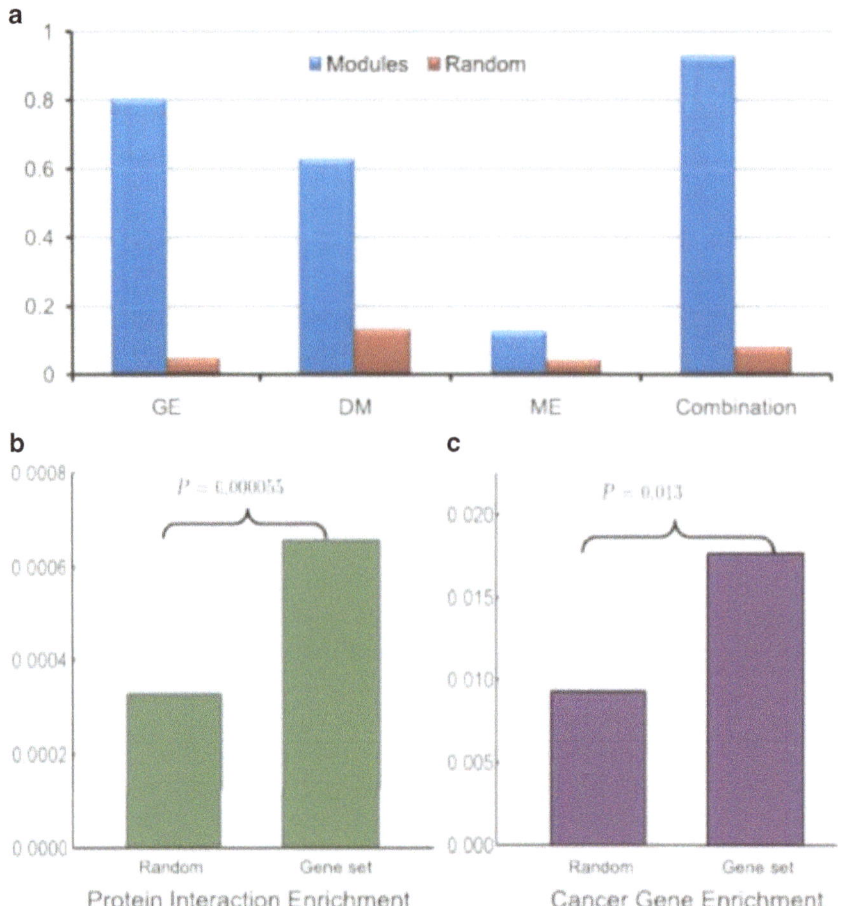

Fig. 2 (**a**) Enrichment ratios of genes related to md-module members in each dimension (GE, DM, or ME) and their combination, with respect to GO biological process terms. For comparison, the mean ratio of functional enrichment for 100 random sets of genes is also plotted. (**b**, **c**) Examples of protein interaction enrichment and cancer gene enrichment, calculated for md-module 173. The *p*-values were determined by the right-tailed Fisher's exact test

both the GE and DM dimensions) are enriched with the cancer gene reference set (p-value <0.05, calculated with the right-tailed Fisher's exact test) (Fig. 2c). The results of the large-scale enrichment analysis support the biological relevance of the regulatory programs detected by our method.

Multidimensional modules can capture multilevel, synchronized disruptions of pathways, thereby facilitating the discovery of malfunctions involving multiple regulatory levels. Thus, this method may aid in the development of drug interventions that can simultaneously correct the effects of various types of dysfunctions.

4.2 Clinical Associations of the md-Modules

The NMF basis vectors (i.e., the columns of W) imply a linkage between each sample and the discovered md-modules. Thus, the NMF framework can discover phenotype-specific md-modules

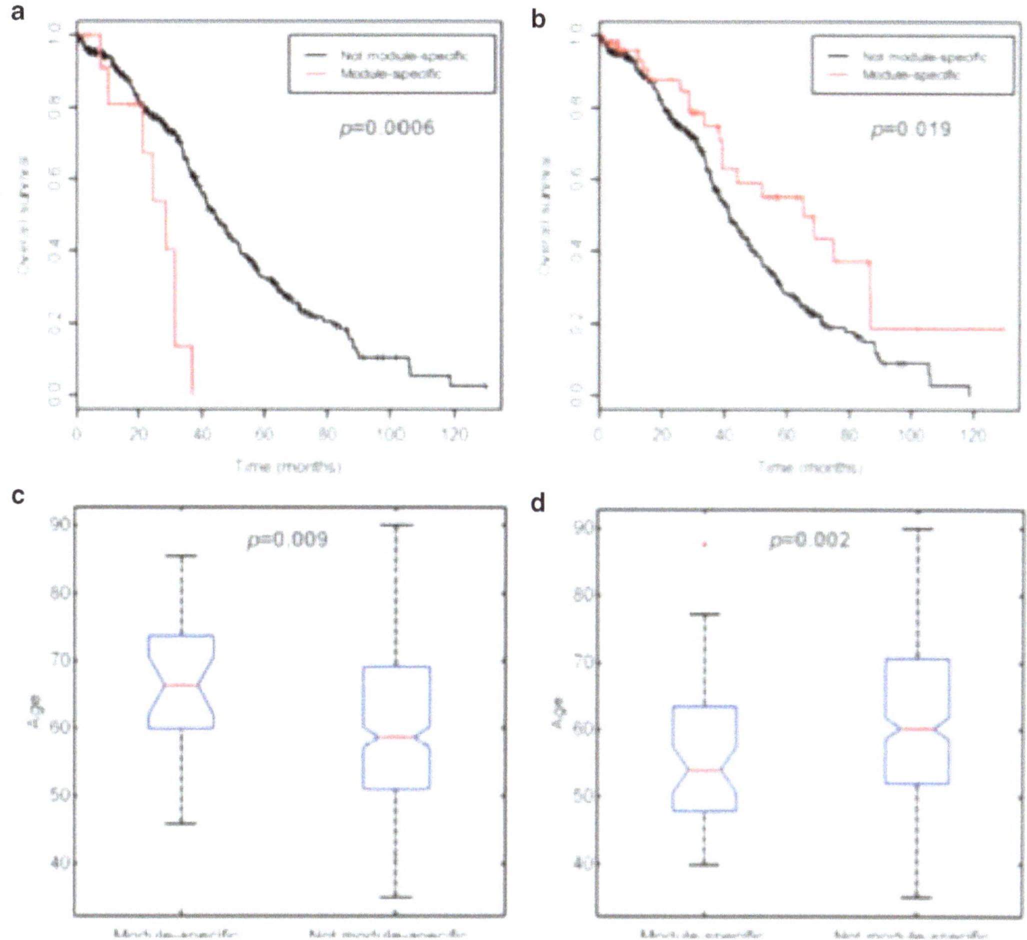

Fig. 3 (**a**) and (**b**) Kaplan–Meier survival analysis for patients associated with module 166 (**a**) or module 3 (**b**), compared to patients outside the module. The *p*-values of their log-rank tests are 0.0006 and 0.019, respectively. The median survivals for patients in module 166 or module 3 versus the outgroup are 26.4 versus 36.1 years and 38.2 versus 33.8 years, respectively. (**c**, **d**) Box-plots for the ages of patients associated with module 28 (**c**) and module 78 (**d**), compared to patients outside the module. The *p*-values of their rank-sum tests are 0.009 and 0.002, respectively. The median ages for patients in module 28 or module 78 versus the out-group are 66.3 versus 58.7 years and 54.1 versus 60.2 years, respectively

by comparing the sample's module relevance to the clinical characterizations of each patient. We compared the survival times of ovarian cancer patients who are strongly associated with a specific md-module to those who are not. We found several md-modules for which the in-group show a significantly shorter or longer median survival time (log-rank test $p<0.05$). For example, 13 patients are strongly associated with md-module 166. They show significantly worse outcomes, with a median survival of 26.4 months compared to 34.1 months for other patients ($p=0.0006$, log-rank test) (Fig. 3a). In fact, these 13 patients show distinctive characteristics in all three dimensions of the md-module. For example, the genes/miRNAs in this module are over/under-expressed, as are the methylation levels of the markers. The module contains numerous cell cycle checkpoint genes (e.g., *BUB1B*, *CENPF*, *MAD2L1*, *CCNB1*, *BUB1*, *CCNA2*, *CHEK1*, and *TTK*), and is significantly enriched in genes from the "nuclear division" functional category (p-value $<10^{-8}$). In another case, the patients in md-module 3 are associated with improved survival, a median of 38.2 months versus 33.8 months in the out-group ($p<0.02$, log-rank test). This module reveals a significant perturbation of the endometrial cancer pathway, with several key genes related to tumorigenesis, e.g., *EGFR*, *CTNNA2*, and *ARAF*.

We identified 20 md-modules whose patients have significantly different age characteristics from patients outside the module. For example, patients in md-module 28 are older (median 66.3 years vs. 58.7 years; $p=0.009$, rank-sum test) (Fig. 3c), and md-module 78 is associated with significantly younger patients (median 54.1 years vs. 60.2 years; $p=0.002$, rank-sum test) (Fig. 3d).

4.3 Application of Semi-supervised Joint NMF

We tested the semi-supervised variant of the NMF framework on a dataset of human miRNA and gene expression profiles from TCGA ovarian cancer samples, using a miRNA–gene interaction network and a gene interaction network as additional constraints in the objective function [10]. We identified 49 human miRNA–gene regulatory co-modules, each with multiple miRNAs and multiple genes. Within each module, the member miRNAs are significantly enriched in miRNA clusters (a miRNA is a set of miRNAs that are closely located in their chromosome), while the member genes are enriched in GO biological processes and KEGG pathways. The overrepresented functional terms of the member genes can potentially be transferred to the member miRNAs, resulting in a functional prediction for miRNAs. We find that the identified co-modules include a significant number of cancer-related genes and miRNAs, and many of these are involved in ovarian cancer. The detected regulatory modules can potentially be used to reconstruct gene regulatory networks, and provide candidates for the experimental validation of miRNA targets. More detailed analysis can be found in [10].

5 Conclusion

With advancing technology, it is becoming standard practice to obtain multi-platform genomic profiles of biological samples, allowing researchers to characterize the complex connections among cellular systems. In this chapter, we have discussed matrix factorization methods for analyzing such data. We demonstrated the method on 385 samples from the TCGA ovarian cancer dataset (comprising gene expression, DNA methylation, and miRNA expression profiles) and other networked data. Matrix factorization detected numerous md-modules corresponding to biologically relevant patterns, and provides some unique insights on the functional and clinical characteristics of these modules [9, 10]. We believe that such modules are a good starting point to uncover the underlying biological mechanisms of gene regulation.

More importantly, matrix factorization methods are equally useful for many other biological problems requiring the integration of several types of data. We observe a trend toward generating multidimensional genomic data (e.g., copy number variation, DNA methylation, histone modification, miRNA expression, and gene expression) on the same set of samples. Our framework would be applicable to these datasets after some preprocessing. At the same time, we are gaining more and more knowledge regarding the connections between different genomic variables. The semi-supervised variant of the joint NMF is a powerful tool for the integration of diverse datasets and known network relationships to discover complex regulatory patterns. For example, Kutalik et al. [5] developed a method to identify gene–drug co-modules using gene expression and drug response data of 60 cancer samples. However, the quality of the modules obtained in this research may suffer from the small number of samples and data noise. By employing our framework, one could incorporate known gene–gene interactions, gene–drug relationships, and even drug–drug similarities to improve the pattern discovery.

Acknowledgements

This project was supported by the National Natural Science Foundation of China, No. 11001256 and No. 61379092 to S.Z., and the National Institutes of Health Grant NHLBIMAPGen HL108634 and a National Science Foundation Career Award 0747475 to X.J.Z.

References

1. McLendon R et al (2008) Comprehensive genomic characterization defines human glioblastoma genes and core pathways. Nature 455:1061–1068
2. Bussey KJ et al (2006) Integrating data on DNA copy number with gene expression levels and drug sensitivities in the NCI-60 cell line panel. Mol Cancer Ther 5: 853–867
3. Zhang W, Zhu J, Schadt EE, Liu JS (2010) A Bayesian partition method for detecting pleiotropic and epistatic eQTL modules. PLoS Comput Biol 6:e1000642
4. Gao F, Foat BC, Bussemaker HJ (2004) Defining transcriptional networks through integrative modeling of mRNA expression and transcription factor binding data. BMC Bioinformatics 5:31
5. Kutalik Z, Beckmann JS, Bergmann S (2008) A modular approach for integrative analysis of large-scale gene-expression and drug-response data. Nat Biotechnol 26:531–539
6. Li W, Zhang S, Liu CC, Zhou XJ (2012) Identifying multi-layer gene regulatory modules from multi-dimensional genomic data. Bioinformatics 28(19):2458–2466

7. Madeira SC, Oliveira AL (2004) Biclustering algorithms for biological data analysis: a survey. IEEE Trans Comput Biol Bioinform 1:24–45
8. Shen R, Olshen AB, Ladanyi M (2009) Integrative clustering of multiple genomic data types using a joint latent variable model with application to breast and lung cancer subtype analysis. Bioinformatics 25:2906–2912
9. Zhang S, Liu CC, Li W, Shen H, Laird PW, Zhou XJ (2012) Discovery of multi-dimensional modules by integrative analysis of cancer genomic data. Nucleic Acids Res 40(19):9379–9391
10. Zhang S, Li Q, Liu J, Zhou XJ (2011) A novel computational framework for simultaneous integration of multiple types of genomic data to identify microRNA-gene regulatory modules. Bioinformatics 27:i401–i409
11. Kim PM, Tidor B (2003) Subsystem identification through dimensionality reduction of large-scale gene expression data. Genome Res 13:1706–1718
12. Brunet JP, Tamayo P, Golub TR, Mesirov JP (2004) Metagenes and molecular pattern discovery using matrix factorization. Proc Natl Acad Sci U S A 101:4164–4169
13. Tamayo P, Scanfeld D, Ebert BL, Gillette MA, Roberts CW, Mesirov JP (2007) Metagene projection for cross-platform, cross-species characterization of global transcriptional states. Proc Natl Acad Sci U S A 104(14):5959–5964
14. Paatero P, Tapper U (1994) Positive matrix factorization: a non-negative factor model with optimal utilization of error estimates of data values. Environmetrics 5:111–126
15. Lee DD, Seung HS (1999) Learning the parts of objects by non-negative matrix factorization. Nature 401:788–791
16. Badea L (2007) Combining gene expression and transcription factor regulation data using simultaneous nonnegative matrix factorization. In proceeding of International Conference on Bioinformatics & Computational Biology, BIOCOMP 2007, Las Vegas Nevada, USA 1: 25–28
17. Kim H, Park H (2007) Sparse non-negative matrix factorizations via alternating non-negativity-constrained least squares for microarray data analysis. Bioinformatics 23:1495–1502
18. Lee D, Seung H (2001) Algorithms for non-negative matrix factorization. Adv Neural Inform Process Syst 13:556–562

Chapter 20

Computational Methods for DNA Copy-Number Analysis of Tumors

Jude Kendall and Alexander Krasnitz

Abstract

Study of DNA copy-number variation is a key part of cancer genomics. With the help of a comprehensive multistep computational procedure described here, copy-number profiles of tumor tissues or individual tumor cells may be generated and interpreted, starting with data acquired by next-generation sequencing. Several of the methods presented are specifically designed to handle cancer-related copy-number profiles. These include accounting for variation of ploidy and distilling somatic copy number alterations from the inherited background.

Key words DNA copy number, Next-generation sequencing, Cancer, Normalization, Segmentation, Single-cell analysis

1 Introduction

Somatic DNA copy number variation is a ubiquitous feature of cancer genomes [1]. The spectrum of alterations is broad and includes both amplifications and deletions, with the genomic extent ranging from less than a single gene to an entire chromosome. Research interest in genome-wide study of these alterations is motivated by their relation to the onset and progression of the disease; by their value as markers for clinical outcome and sensitivity to treatment [2]; and, more recently, by their utility in reconstructing the genealogy of tumor cells [3]. In addition to comparative genomic hybridization, next-generation sequencing (NGS) is being increasingly employed as a method for copy-number data acquisition. The obvious advantage of NGS over array-based technology is its ability to provide truly genome-wide coverage, not restricted to array probes.

Here, we summarize methodologies we have been using for in silico copy-number analysis, with the focus on the analysis of NGS-derived data. Starting with a set of sequencing reads, a succession of computational tools produces first noisy, then piecewise-constant

Narendra Wajapeyee (ed.), *Cancer Genomics and Proteomics: Methods and Protocols*, Methods in Molecular Biology, vol. 1176, DOI 10.1007/978-1-4939-0992-6_20, © Springer Science+Business Media New York 2014

estimates of copy number profiles of tumor tissues or individual tumor cells, These are further interpreted to identify copy-number alteration events.

2 Methods

There are many methods for estimating DNA copy number genome-wide. Recent large-scale studies have used various microarray and high-throughput sequencing methods [1, 3]. In all cases the idea is to measure a quantity that correlates predictably with DNA copy number. In the case of microarrays the fluorescent intensity of spots on the array is higher for higher copy number. For high-throughput sequencing the depth of coverage correlates with copy number. Also, paired end or mate pair sequencing can be used to find deletions and amplifications. We describe here data analysis methodology to infer DNA copy number alterations from high-throughput sequence data. Although customized for processing data generated by Illumina sequencing machines, this methodology can be readily adapted to handle output from other sequencing platforms. A reader interested in the details of sequencing technology should consult the Illumina Web site [4].

Important aspects of our methodology depend on whether DNA being analyzed originated in a single cell or in bulk tissue. In case of the former, we take advantage of the fact that copy number is integer everywhere in the single-cell genome. In case of the latter, the observed copy number profile is an average over cells representing a variety of clonal populations in the tumor, admixed with normal tissues, and we must account for this additional source of variability.

Our focus here is exclusively on in silico analysis of individual copy number profiles. Experimental techniques underlying this analysis, in particular DNA profiling of single cells, are described in detail in separate publications [3, 5]. On the other hand, single-profile analysis often is followed up by examination of multiple copy-number profiles, to infer regions of copy number aberration recurrent in tumors from multiple patients [6] or to infer clonal structure and tumor evolution in the case of single cells from one tumor or patient [3]. Methodology of this multi-profile analysis is also presented in detail in a separate publication [7].

The main idea used here in inferring DNA copy number from high-throughput sequence data is that the read depth in a region of the genome is roughly proportional to the DNA copy number. Deviations from this proportionality are mainly due to two factors. First, Illumina sequencing machines produce short read data, generally 50–150 bases long. A sizable fraction (10–15 %) of these short reads cannot be uniquely localized in the genome. Secondly, there are sequence-dependent biases affecting the read depth, including, but likely not limited to, those related to PCR amplification, adapter ligation and spot forming efficiency on the Illumina flow cell surface. Local GC content of the genome affects all these steps and is correlated with large biases, variable from sample to sample, but can be easily normalized (*see* **Note 1**).

The data analysis consists of the following steps,

1. Partition the genome into bins with an expected equal number of mappable positions.
2. Count the number of reads mapping to each genome bin.
3. Remove from subsequent analysis bins that have consistently high read counts.
4. Normalize the read counts based on GC content of each bin.
5. Estimate the underlying piecewise-constant profile by segmentation.
6. If analyzing single cell data, for each segment estimate an integer copy number.
7. Center the profiles and determine segments with altered copy number.
8. If analyzing bulk DNA, determine segments with altered copy number by comparison with profiles of cancer-free tissues.
9. Mask inherited copy number alterations.
10. Identify copy number altering events.

These individual steps are organized into a workflow, as shown in Fig. 1. Software code samples for **steps 1–5** are given in [5]. Software packages for **steps 7–9** are publicly available from the Comprehensive R Archive Network (CRAN) at the following Web locations: http://cran.us.r-project.org/web/packages/ParDNAcopy/ and http://cran.us.r-project.org/web/packages/CNprep/.

2.1 Partition the Genome into Bins with an Expected Equal Number of Mappable Positions

The Illumina sequencing machines produce 50–150 bases of sequence data from the ends of DNA molecules. This can be done by sequencing just one end of each molecule, called single read sequencing, or by sequencing both ends of each molecule, called paired end sequencing. We count the number of molecules that map to each genome bin, so we typically use single read sequencing. The sequencing machines output a data file in FASTQ format with four lines for each read, illustrated in Fig. 2a. Line 1 is a read ID, line 2 is a sequence consisting of A, C, G, T, or N, line 3 is a plus sign, and line 4 is a quality score for each base [8]. The quality score is an ASCII-encoded phred quality score [9]. Mapping these reads to a reference genome consists in finding the 100 base long sequence in the reference genome that most closely matches the sequence read. There are usually base calling errors within each

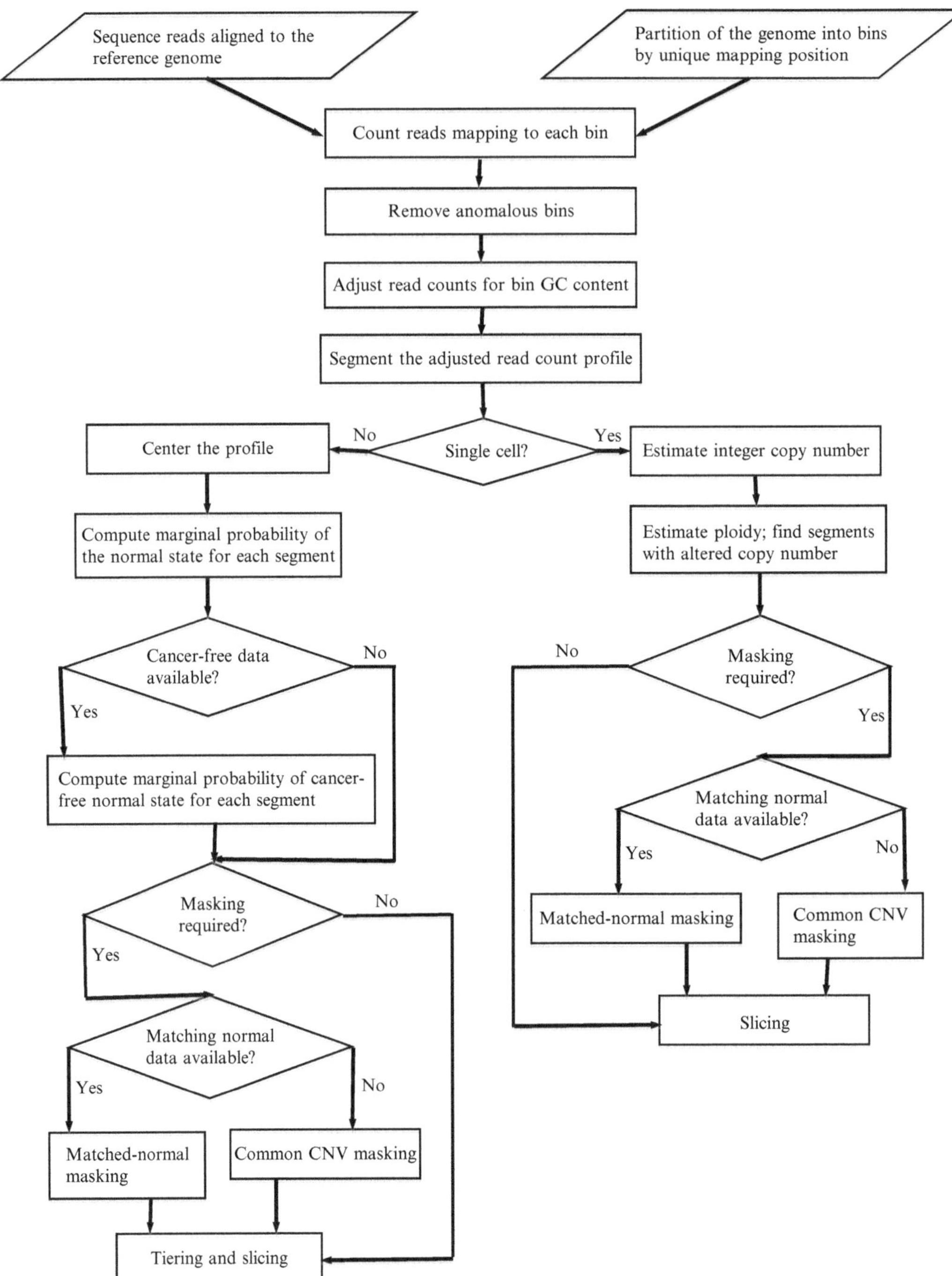

Fig. 1 A flowchart for the computational procedure described in Subheading 2

A

```
Read data from sequencing machine in fastq format.
@GA200001:79:FC64N77AAXX:7:69:10027:15968 1:N:0:
GACAAAATTTCATTCTTGTTGACCAGGTTGGAGTGCAA
+
DDDDBDD>DDDDDDDDDDDDD<DDBDDDDDDD>9,B>B
```

B

```
Read data after mapping in SAM format.
GA200001:79:FC64N77AAXX:7:69:10027:15968   16      chr10   77560   255
38M     *       0       0       TTGCACTCCAACCTGGTCAACAAGAATGAAATTTTGTC
B>B,9>DDDDDDDBDD<DDDDDDDDDDDDD>DDBDDDD  XA:i:0  MD:Z:38 NM:i:0
```

Fig. 2 Illumina sequencing data format. (**a**) For each spot on the flowcell surface the Illumina sequencing machines output 4 lines of data in FASTQ format. *Line 1* is a read ID, *line 2* lists the bases called, *line 3* is a plus sign and *line 4* is the phred quality score for each base, ASCII-encoded. (**b**) A SAM file format of an alignment of a sequence read to the human genome. There is one tab-delimited line for each read aligned, showing the read ID, read sequence and quality scores and the alignment position in the reference genome

read, especially in the low quality base calls occurring at the end of the read. An alignment with the fewest mismatches to the reference genome is found, and if no other location in the genome aligns with the read with the same number of mismatches, then the read is considered to be uniquely mapped to the reference genome. There are many software packages available for short read alignment. In this analysis we use Bowtie [10]. Bowtie takes as input the sequence reads in FASTQ format and a reference genome sequence and outputs alignments in SAM file format [11]. Figure 2b is an example of an alignment in SAM format. There is one tab-delimited line for each alignment. The fields of interest to us are: field 1, the read ID; field 2, a mapping code, which for single read mappings is 0 for an alignment on the plus strand and 16 for an alignment on the minus strand; and fields 3 and 4, the chromosome and chromosome position to which the read aligns.

Not every 50-base sequence in the human genome is unique to just one location (*see* **Note 2**). For example, the sequence consisting of 50 A-s matches over 5000 places in the GRCh37 reference human genome. For such a read we cannot know from where in the genome it was sampled. In a typical sequencing protocol DNA is extracted from one or many cells and in effect randomly sampled and sequenced. The DNA library will consist of billions of short DNA fragments, typically about 300 bases long. Output from a sequencing run can contain from 10 to 300 million reads. Samples can be indexed and sequenced together in a library giving as few as one million molecules per sample (*see* **Note 3**). The simplest read depth data analysis could consist of partitioning the genome into for example 100,000 bins of equal length of about 30,000 bases in the case of the human genome. But due to the existence of a large amount of repetitive sequence in the human genome, from transposons, microsatellites, and segmental duplications, some of these bins will have very few 50 base sequences that are unique in the genome.

In order to compensate for these biases we map every 50 base long sequence (50-mer) in the reference genome. For each of these sequences we know the chromosome and chromosome position. Start positions for 50-mers that are unique in the genome are called uniquely mapping positions. We then partition the genome into 20,000 bins each with the same number of uniquely mapping positions, plus or minus 1 to handle rounding to integers (*see* **Note 4**). For 20,000 bins and the GRCh37 version of the reference genome the median bin size is about 120,000 uniquely mapping positions. The base count varies from bin to bin, and some bins near centromeres may span many megabases.

2.2 Count the Number of Reads Mapping to Each Genome Bin

For each uniquely mapped sequence read we know the chromosome and chromosome position of its alignment. We take the starting position of each alignment, find the bin in which it is contained and increment the read count for that bin. The result of counting each read is a file as illustrated in Table 1. The first column is the bin number, the second column is the chromosome of the bin, the third column is the starting position of the bin along the chromosome (the bin ends at the position just before the start of the subsequent bin), the fourth column is the bin count or the number of reads uniquely mapping to a position within the bin boundaries.

2.3 Remove Anomalous Bins

There are dispersed regions in the genome that accumulate very high read depth compared to expected read depth based on the number of reads, for example, the bins surrounding the centromere

Table 1
A bin count data table for the first 5 bins of chromosome 1 in a genome partitioned into 20,000 bins

Bin	Chrom	Chrompos	Bincount	Ratio	GC.content	Lowratio	Seg.mean	Ratio.q	Seg.q
1	1	0	53	1.09	0.45	1.03	1.00	3.15	3.06
2	1	892036	31	0.65	0.63	0.82	1.00	2.50	3.06
3	1	1033528	49	1.01	0.60	1.18	1.00	3.60	3.06
4	1	1172212	59	1.21	0.62	1.53	1.00	4.67	3.06
5	1	1311902	56	1.15	0.57	1.23	1.00	3.74	3.06

Each row corresponds to a bin. The columns, from left to right, tabulate the ordinal bin number ("bin"), the chromosome number ("chrom"), the bin start position in the chromosome ("chrompos"), the read count in the bin ("bincount"), the read count normalized by the mean read count per bin ("ratio"), the proportion of G or C in the bin ("gc.content"), the normalized read count with the GC bias removed ("lowratio"), the segmented normalized read count ("seg.mean"), the "lowratio" column multiplied the optimal M (cf Methods, 2.6, "ratio.q"), the optimal MS (cf Methods, 2.6, "seg.q")

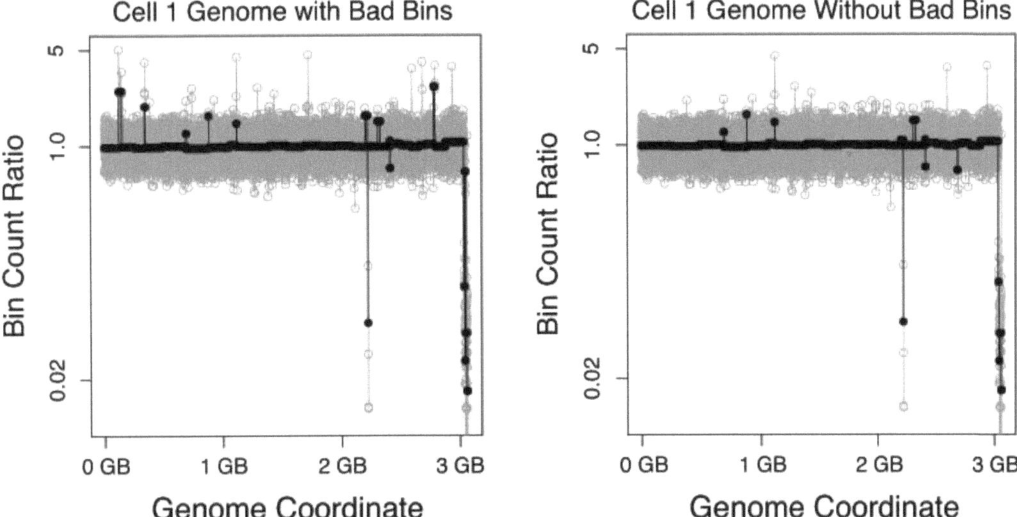

Fig. 3 Identification and removal of anomalous bins. The *left panel* shows a segmented copy-number profile of a diploid single-cell genome with all 20,000 bins included in the data analysis. The *right panel* shows a segmented profile of the same cell after removing from analysis bins at centromeres with consistently high read counts in many diploid cells. In this case 38 bins are removed leaving 19,962 bins. The normalized bin counts prior to segmentation are shown in *gray* in both panels

on chromosome 1. These bins consistently, across the vast majority of the profiles, have 5 to 10 times as many reads as the median bin count across the genome. A succession of such bins would be mistakenly identified as a region of copy number gain. An example is shown in Fig. 3 where these bins are referred to as "bad bins." Anomalously high read counts are observed in the very large bins consisting of mostly repetitive DNA sequence. For example, there are over a million Alu elements in the human genome. The older copies are mutated over time. Short read mapping, base mutations in the reference genome, base-calling errors in the sequence data, and errors in the reference genome in highly repetitive regions combine to make mapping in these regions less reliable than is generally the case. We have a list of bins empirically determined to consistently accumulate an unusually high number of mappings. These bins are removed from subsequent data analysis (*see* **Note 5**).

2.4 Normalize the Read Counts Based on the GC Content of Each Bin

There are large biases in read count correlated with GC content. The result is that the read counts for each bin, rather than being directly proportional to DNA copy number, are distorted. This phenomenon is illustrated in Fig. 4 which shows a scatter plot of bin read count against the bin GC content, defined as the share of G or C in the reference sequence of the bin. That is, bins with high GC content tend to have lower bin counts than bins with an average GC content. Importantly, the GC bias varies from sample

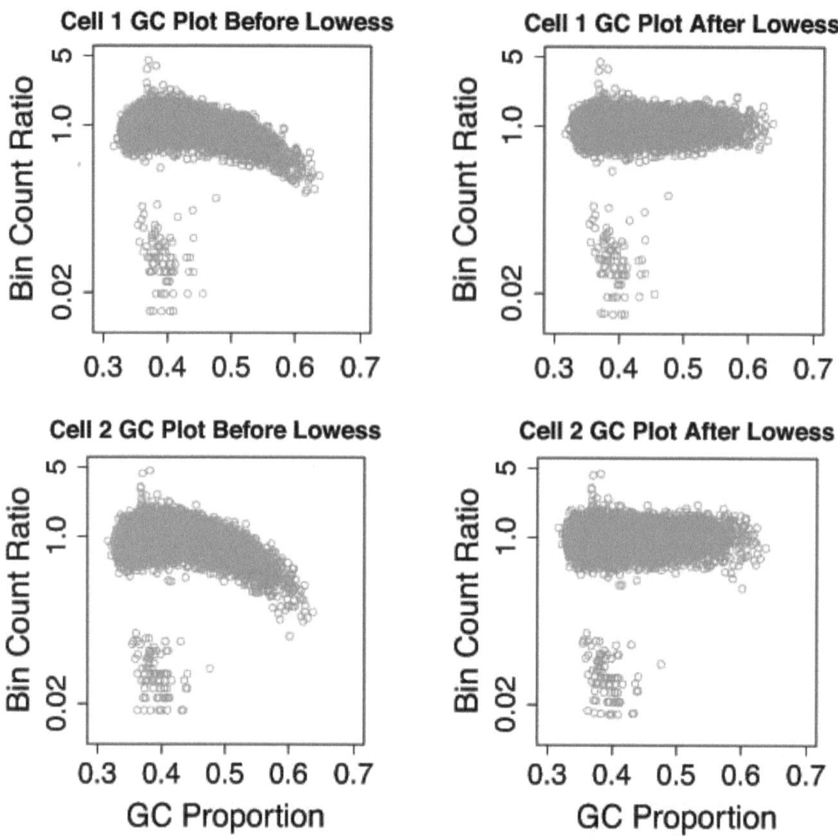

Fig. 4 Sample-dependent GC bias and its removal. The two plots on the left show read count versus GC content for the genome bins in two different diploid cells from a breast cancer patient. The bins with very low read count are from the Y chromosome. These plots illustrate that the GC bias differs between samples and must be adjusted for separately for each sample. The right two plots show the same data after GC bias removal

to sample, meaning that whereas variation of mappability across the genome could be assessed once and adjusted for by making bins of variable length, it is necessary to adjust the read counts for GC content in each sample individually. To this end we fit a Lowess curve to the GC content by bin count data and subtract from each bin read count its respective Lowess estimate. As can be seen in Fig. 2, the bias is thereby dramatically reduced. This correction is analogous to using Lowess fit for the normalization of intensity-dependent dye bias in two-color microarray data [12]. The original and GC-adjusted read count values are also compared for a small number of bins in Table 1.

2.5 Estimate the Underlying Piecewise-Constant Profile by Segmentation

The bin read count as a function of the genomic position represents the copy number of the sample superimposed with observational noise. The copy number itself is a piecewise-constant function of the genomic position, resulting from copy number gains and losses, both inherited and caused by cancer.

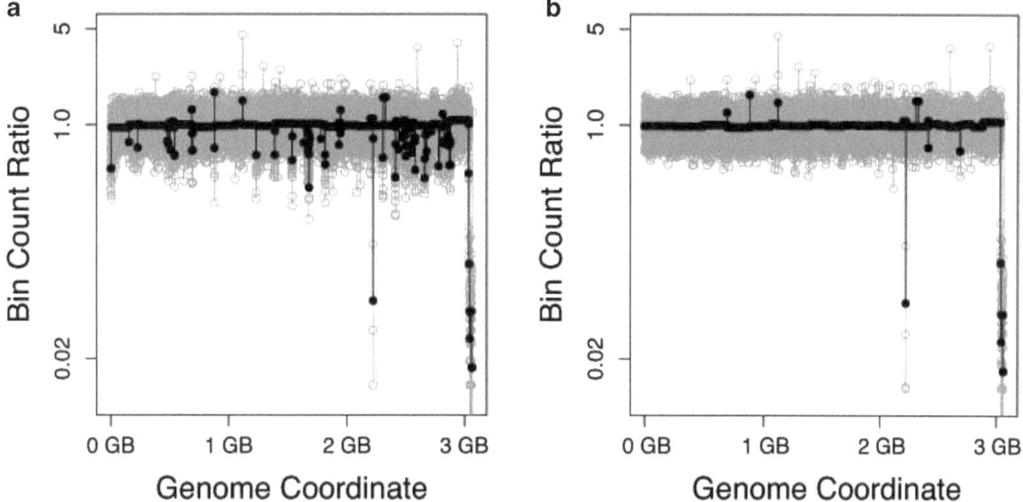

Fig. 5 The effect of GC bias removal on segmentation. (**a**) A segmented profile for a diploid cell prior to GC bias removal. Multiple spurious deletions are seen in regions of high GC content, frequently located near telomeres. (**b**) A segmented profile for the same cell after GC normalization. The copy number zero region on the far right of both panels is the Y chromosome. The normalized bin counts prior to segmentation are shown in *gray* in both panels

Regions of copy-number variation vary in length from whole chromosomes and chromosome arms down to single genes. Our ability to find these regions depends both on the level of noise in the data and on the sampling rate, i.e., the number of bins per base pair. Usually a variant region cannot be reliably detected unless it spans multiple bins. A procedure for estimating the underlying piecewise-constant copy-number profile from noisy bin-count data is called segmentation. We use the DNAcopy segmentation algorithm for this purpose [13]. In single cells each of these segments corresponds to a region of chromosome with an integer number of copies. In DNA samples extracted from many cells the copy number in each of these regions represents an average across a population of cells. The segmented profile is compared with the original noisy data genome-wide in Fig. 5 and for a small number of bins in Table 1. Figure 5 also illustrates the importance of bias removal prior to segmentation.

2.6 If Analyzing Single-Cell Data, for Each Segment Estimate an Integer Copy Number

As indicated above, when analyzing data from a single cell we know that the copy number must be an integer at any point in the genome. This means that if the segmented values accurately represent the copy number in each bin then the genome-wide distribution of these values should be multimodal and sharply peaked, with each mode representing an integer copy number value present in the genome. With this expectation we estimate the copy number genome-wide, assuming in addition that (a) the segmented values are approximately proportional to the copy number and

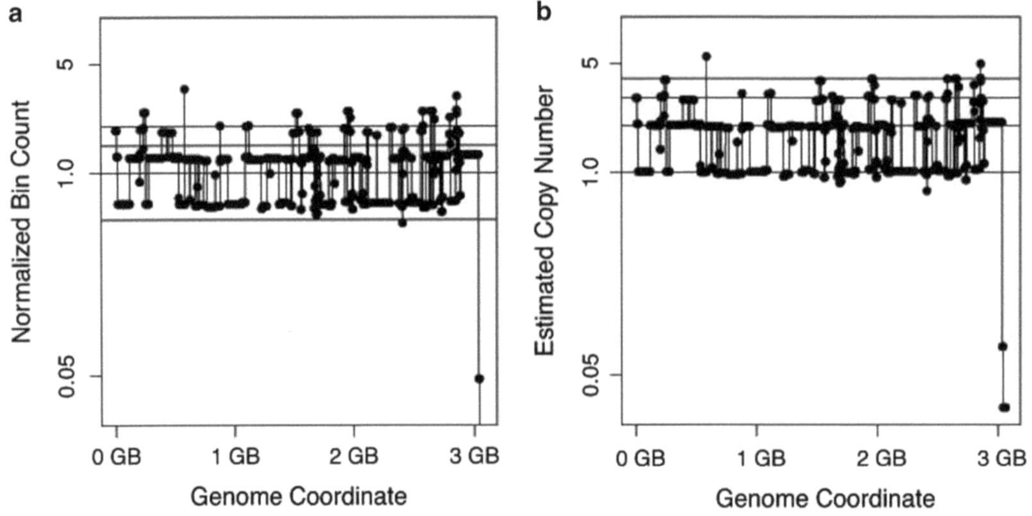

Fig. 6 Integer copy number estimation. (**a**) A segmented profile for a hypo-diploid tumor cell with DNA content of approximately 0.85 that of a diploid genome. The segmented values are normalized so that their genome-wide mean is 1. The *horizontal lines* are drawn at the values of 0.5, 1.0, 1.5, and 2.0 of the vertical coordinate. (**b**) The integer-valued copy-number profile of the same cell, determined by minimization of the mean squared rounding error. The *horizontal lines* are drawn at copy number values of 1, 2, 3, and 4

(b) the genome-wide mean of the copy number lies in the 1.5–5.5 range. The estimate is obtained by least-squares fit [14]: denoting the nearest integer by [], the genome-wide mean by ⟨⟩ and the segmented values by S, we minimize the mean squared rounding error $\langle (MS\text{-}[MS])^2 \rangle$ with respect to multiplier M, subject to the condition $1.5 \leq \langle MS \rangle \leq 5.5$. The estimated copy number is then given, for each bin, by the corresponding $[MS]$ computed at the optimal value of M. The first additional assumption we make is justified a posteriori: the mean squared rounding error is small for the vast majority of single cells we have analyzed. Armed with this experience, we use this error for quality control: copy number profiles with large errors either may originate from erroneous pooling of DNA from multiple cells, or derive from parts of cells, or represent degraded DNA (*see* **Note 6**). Our second additional assumption is justified by observation: cell ploidy, as measured by flow cytometry, very rarely lies outside the 1.5–5.5 range. Indeed, we find the optimal value of $\langle MS \rangle$ to lie in the interior of this range in a vast majority of cases.

The copy number estimation step is illustrated in Fig. 6. The left panel of the figure shows a segmented genome-wide profile of a single cell from the hypo-diploid cell population [3]. It is evident that the genome is predominantly, and nearly evenly, distributed between two values of the copy number, one of which is roughly twice the other. These two values are estimated as described to be 2 and 1 (right panel). Estimated values of MS are also listed in Table 1.

2.7 Center the Profiles and Determine Segments with Altered Copy Number

Tumor cells commonly undergo whole-genome copy number gains and losses, resulting in changed ploidy. It is important to be aware of these changes in ploidy in order to correctly infer individual copy number alterations in single-cell profiles. For example, copy number of 2 would be a loss in a triploid but not in a diploid genome. To estimate the ploidy of a cell, we first find, for each chromosome other than X or Y, the integer copy number where most of the chromosome's length resides, and then take the most frequent of these integers among the chromosomes considered as the ploidy of the cell. Copy number changes in individual segments are computed relative to ploidy.

In the analysis of bulk tumor DNA cells comprising the sample need not all have the same ploidy or harbor the same copy number alterations. As a result, segments comprising the profile need not differ by integer copy number, and it is not obvious what value of the copy number estimate corresponds to the normal state of the cells, relative to which somatic alterations occur. We therefore adopted the following multistep process to infer the likely normal state and identify segments with significant copy number deviation from that state. First, a bootstrap sample is generated from the log normalized bin count data for the median value of each segment. The bootstrap sample size is proportional to the number of bins in the segment (*see* **Note 7**). These samples are joined, and the result is modeled as a normal mixture using expectation maximization (EM) algorithm [15]. We usually allow the mixture components to both have differing means and standard deviations. The optimal number of components for the mixture is chosen based on the Bayesian information criterion (BIC). The algorithm is re-initialized at random 10 times, and the result with the highest BIC is chosen as final. This result contains, in addition to the component means and standard deviations, the component identities for all values in the sample. The sample is therefore divided into clusters depending on the component label. Next, we seek the maximal overlap of any two components. If that overlap exceeds a predetermined value (typically 0.25–0.4), the two clusters are joined. This search and joining of the most overlapping pair of clusters is repeated until no two of the remaining clusters overlap by more than the preset value. At this point in the process we seek the cluster whose median value is nearest 0. If that cluster represents more that a certain predetermined portion of the genome, it is declared a normal-state cluster (*see* **Note 8**). Otherwise, the two clusters with medians nearest 0 are joined, a process repeated until the result represents at least the required portion of the genome. Once the normal-state cluster is determined, its median is taken as the estimate of the normal-state log segmented value. We then estimate and tabulate, for each segment, its marginal probability to belong to that cluster. To that end, the segment median is bootstrap-sampled a fixed number of times (typically 20–50), and the marginal probability to belong to the normal-state cluster is averaged over the sample values. Segments for which this probability is low are flagged as likely to be amplified or deleted, depending on the sign of their deviation from the normal state. This criterion may be used alone or in combination with a criterion based on comparison of the tumor-derived data to those derived from cancer-free tissues.

2.8 If Analyzing Bulk DNA, Determine Segments with Altered Copy Number by Comparison with Profiles of Cancer-Free Tissues

If a representative set of copy number profiles is available for cancer-free tissues, it can be used to identify segments with significant copy number deviation from the norm in tumor-related profiles. For this identification to be meaningful it is essential for the tumor-related and cancer-free DNA to have undergone the same preservation, extraction and sequencing procedures. If this is the case, we remove from the cancer-free data all segments whose copy number is likely to deviate from 2. These are short segments, usually defined as shorter than 5 Mb, and segments in the chromosomes X and Y. For the remaining cancer-free data we determine, separately for each profile, the genome-wide median for the log of segmented values and tabulate, for each segment, its deviation from the median. Next, tumor-derived segments are compared one by one to the joint set of long, autosomal cancer-free segments. This comparison yields, for each tumor-derived segment, the fraction of the total length of the cancer-free segment set for which the log segmented value deviation from the norm exceeds that of the tumor-derived segment. This fraction is small for tumor-derived segments with altered copy number and may therefore be used to identify amplified or deleted segments, alone or in combination with the clustering-based criterion described in the previous subsection.

2.9 Mask Inherited Copy Number Alterations

Some of the amplified or deleted segments identified in tumor-derived copy number profiles are not somatic and specific to the tumor tissue or cell but inherited and common to all cells of the organism. These inherited alterations may be masked or retained in the profile, depending on the final goal of the analysis. In particular, if the goal is to determine somatic alterations across multiple tumors, it is advisable to mask the inherited alterations. If, on the other hand, the goal is to reconstruct the genealogy or the population structure of cells within a tumor, this step may not be necessary, as such inherited alterations are shared by all the cells under investigation and will have no impact on the analysis downstream.

If masking is necessary, its implementation and results depend on the availability of the copy number profile of a matching normal tissue. If available, a matching profile makes it possible to mask all inherited alterations, independent of their frequency in the population. We then proceed as follows. First, we identify amplified (deleted) segments in the tumor profile that are matched

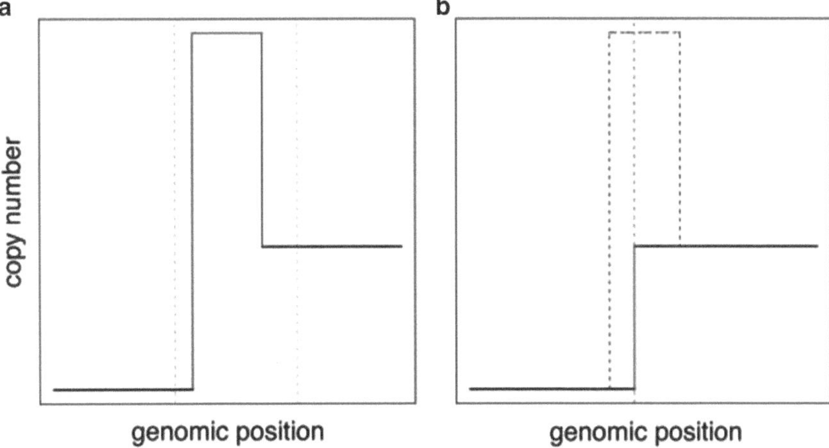

Fig. 7 Masking of inherited copy-number variation. (**a**) A portion of a segmented profile. An amplified segment (shown in *red*) is overlapped by an inherited amplified region (a region between the *green dashed lines*). (**b**) A position, shown by a *green dashed line*, is chosen at random within the amplified segment. The neighboring segments are extended to that position (Color figure online)

closely by amplified (deleted) segments in the normal profile. The matching is quantified as the fraction of the tumor-derived segment covered by the segment from the normal profile. We usually require this measure to exceed 0.6 to trigger masking. The segment slated for masking is then removed as described in Fig. 7: a point is chosen at random within the segment and neighboring segments are extended to the chosen point.

If a matching profile is not available, we must rely on population-wide inherited copy number variation data to generate a mask. The mask is defined by two frequency thresholds, the high threshold T, $0<T<1$, and the low threshold t, $0<t\leq T$. For a given type of alteration (amplification or deletion) we then identify all contiguous regions of the genome where the alteration frequency exceeds T. Each such region is embedded in a (usually broader) region where the frequency exceeds t. The latter region is added to the mask. The mask is thus a set of genomic intervals. If an altered segment of the same type in the tumor is matched by one of these intervals, it is masked as described above. By definition, there will be no masking of copy number alterations whose frequency in the population does not exceed t.

2.10 Identify Copy Number Altering Events

Segments observed in a copy-number profile do not necessarily coincide with the copy-number altering events that gave rise to the observation. However, these events may be reconstructed from the observed profile under certain assumptions. Our reconstruction procedure, called slicing, is especially straightforward for single-cell profiles, where all copy-number values are integer.

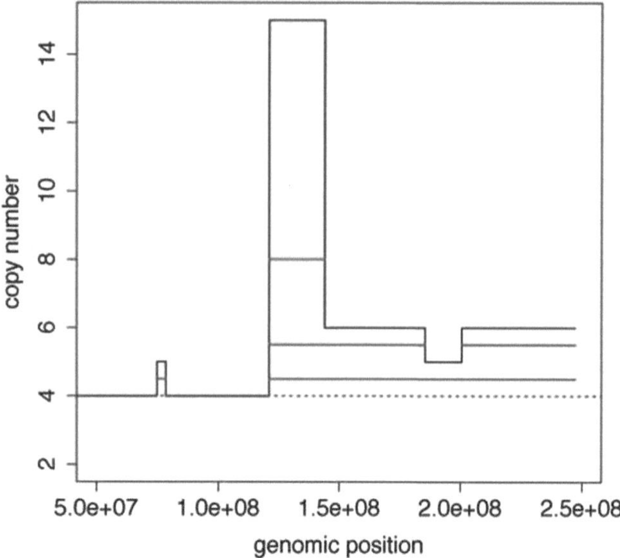

Fig. 8 Slicing. A portion of the integer copy number profile of chromosome 1 in a cell is sliced into four unique interval events (in *red*). The ploidy value for this cell is 4 (shown by a *dashed line*) (Color figure online)

The assumptions underlying slicing are that (a) all copy-number events that have occurred in the ancestral history of the cell are observable in its copy-number profile, (b) there is no event overlap involving both amplification and deletion events and (c) the only allowed form of event overlap is nesting. The procedure consists of first identifying event blocks in the profile. An event block is a succession of segments, all of the same type (amplification or deletion), either ending at a segment that is not of that type or at a chromosome boundary. Consistent with the assumptions (a–b), the events are then found by maximally extending each segment within the profile (Fig. 8).

For the slicing procedure as described to work events in the event block must all correspond to integer copy-number alterations. This property makes it easy to determine which segments form a copy-number event. However, this property does not hold for copy-number profiles obtained from bulk tumor DNA. Application of slicing to such profiles requires the following preliminary step we call tiering [16]. The tiers are formed by first selecting a group of seed segments in the block with a maximal total length subject to a condition that any two segment medians in the group are different with 95 % confidence. The tiers in a block are numbered in the order of increasing seed median. The remaining segments are each assigned to the tier with the nearest seed median. Slicing is then carried out in the block, with the tier being the analog of the integer copy number in the single-cell case.

3 Notes

1. For single cell analyses in our lab we use the Sigma Genomeplex whole genome amplification (WGA) kit. Rubicon's Picoplex WGA kit has different biases and must therefore be normalized using different methods to get optimal results. A similar adaptation is required for any other method of DNA sample preparation.
2. Bin boundaries are set giving bins with equal mappability of uniquely mappable 50-base sequences. Using barcoded (indexed) WGA DNA samples there are 38 bases of barcode and primer to remove from the 5′ end of each read. For 76 base reads this leaves 38 mappable bases and for 100 base reads 62. We have found that bins based on mappability of 50 base reads work similarly enough for 38 base and 62 base reads that these bin boundaries can be used in both cases. For a given single study with different samples sequenced at different lengths the longer reads are truncated so that the same number of bases are mapped for all samples. Bin boundaries can be computed for the desired mapped read length.
3. In our studies of tumor heterogeneity we are interested in profiling the genomes of as many cells from a tumor as we can afford, so we try to get a usable genome copy number profile with as few reads as possible. For aneuploid solid tumors we have found that splitting the genome into 5,000 bins and generating 250,000 reads per cell is useful. This corresponds to a read depth of 1 across 5 % of the human genome. This approach is different from that of most published sequence data analyses, where the goal is to get as much coverage of the genome as possible and high read depth in order to accurately determine point mutations and SNPs.
4. For the examples in this paper we show data from the human genome partitioned into 20,000 bins. This gives good copy number profiles with about 1 million reads per cell. If the purpose of a study is to sample as many cells as possible from a tumor at low cost, fewer bins can be used. This might be the goal in a study of tumor heterogeneity. In a study with some interest in finding amplified or deleted genes more sequence data can be obtained by sequencing fewer samples per flowcell lane and partitioning the genome into smaller bins. This would allow the detection of smaller regions of variation such as a single gene at copy number zero, or a small sub-megabase region amplified to many copies.
5. We found that for a 20,000-bin partition of the genome 38 bins consistently had anomalously high read counts and had to be removed. The reads in these bins are not uniformly distributed.

There are sparsely distributed regions of high read density of about 1,000 bases in size. An alternative to removing these bins would be to find and mask these regions throughout the genome. Preliminary results show that this approach can reduce read count variability throughout the genome in addition to removing the false positive amplification segments around the centromeres. Work is in progress on this alternative method.

6. The method to convert the segmented values to a copy number estimate described in our earlier [3] and in more detail in our more recent [5] publications is different than the method described here. There are two reasons to prefer the new method. First, unlike the earlier method, it is applicable to cancer-free cells, with copy number of 2 nearly everywhere in the genome. Secondly, the minimal value of the mean squared rounding error indicates the quality of the estimate and can be used for quality control.
7. This proportionality coefficient typically is in the 0.02–0.05 range.
8. The minimal portion of the genome to be represented by the normal-state cluster is tuned depending on the cancer type. In breast cancer the expectation generally is that no less than 40–50 % of the genome retains its original copy number; for ovarian cancer that expectation may be as low as 15–20 %.

References

1. Beroukhim R, Mermel CH, Porter D, Wei G, Raychaudhuri S, Donovan J, Barretina J, Boehm JS, Dobson J, Urashima M, Mc Henry KT, Pinchback RM, Ligon AH, Cho YJ, Haery L, Greulich H, Reich M, Winckler W, Lawrence MS, Weir BA, Tanaka KE, Chiang DY, Bass AJ, Loo A, Hoffman C, Prensner J, Liefeld T, Gao Q, Yecies D, Signoretti S, Maher E, Kaye FJ, Sasaki H, Tepper JE, Fletcher JA, Tabernero J, Baselga J, Tsao MS, Demichelis F, Rubin MA, Janne PA, Daly MJ, Nucera C, Levine RL, Ebert BL, Gabriel S, Rustgi AK, Antonescu CR, Ladanyi M, Letai A, Garraway LA, Loda M, Beer DG, True LD, Okamoto A, Pomeroy SL, Singer S, Golub TR, Lander ES, Getz G, Sellers WR, Meyerson M (2010) The landscape of somatic copy-number alteration across human cancers. Nature 463:899–905
2. Hicks J, Krasnitz A, Lakshmi B, Navin NE, Riggs M, Leibu E, Esposito D, Alexander J, Troge J, Grubor V, Yoon S, Wigler M, Ye K, Borresen-Dale AL, Naume B, Schlicting E, Norton L, Hagerstrom T, Skoog L, Auer G, Maner S, Lundin P, Zetterberg A (2006) Novel patterns of genome rearrangement and their association with survival in breast cancer. Genome Res 16:1465–1479
3. Navin N, Kendall J, Troge J, Andrews P, Rodgers L, McIndoo J, Cook K, Stepansky A, Levy D, Esposito D, Muthuswamy L, Krasnitz A, McCombie WR, Hicks J, Wigler M (2011) Tumour evolution inferred by single-cell sequencing. Nature 472:90–94
4. Illumina sequencing technology. http://www.illumina.com/technology/next-generation-sequencing.ilmn. Accessed 20 May 2013
5. Baslan T, Kendall J, Rodgers L, Cox H, Riggs M, Stepansky A, Troge J, Ravi K, Esposito D, Lakshmi B, Wigler M, Navin N, Hicks J (2012) Genome-wide copy number analysis of single cells. Nat Protoc 7:1024–1041
6. Bric A, Miething C, Bialucha CU, Scuoppo C, Zender L, Krasnitz A, Xuan ZY, Zuber J, Wigler M, Hicks J, McCombie RW, Hemann MT, Hannon GJ, Powers S, Lowe SW (2009) Functional identification of tumor-suppressor genes through an in vivo RNA interference

screen in a mouse lymphoma model. Cancer Cell 16:324–335
7. Krasnitz A, Sun G, Andrews P, Wigler M (2013) Target inference from collections of genomic intervals. Proc Natl Acad Sci 110(25): E2271–E2278
8. Cock PJA, Fields CJ, Goto N, Heuer ML, Rice PM (2010) The Sanger FASTQ file format for sequences with quality scores, and the Solexa/Illumina FASTQ variants. Nucleic Acids Res 38:1767–1771
9. Ewing B, Green P (1998) Base-calling of automated sequencer traces using phred. II. Error probabilities. Genome Res 8:186–194
10. Langmead B, Trapnell C, Pop M, Salzberg SL (2009) Ultrafast and memory-efficient alignment of short DNA sequences to the human genome. Genome Biol 10(3):R25
11. Li H, Handsaker B, Wysoker A, Fennell T, Ruan J, Homer N, Marth G, Abecasis G, Durbin R, Proc GPD (2009) The sequence alignment/map format and SAMtools. Bioinformatics 25:2078–2079
12. Yang YH, Dudoit S, Luu P, Lin DM, Peng V, Ngai J, Speed TP (2002) Normalization for cDNA microarray data: a robust composite method addressing single and multiple slide systematic variation. Nucleic Acids Res 30(4):e15
13. Olshen AB, Venkatraman ES, Lucito R, Wigler M (2004) Circular binary segmentation for the analysis of array-based DNA copy number data. Biostatistics 5:557–572
14. Wigler M (2013) Private communication
15. Fraley C, Raftery AE (2002) Model-based clustering, discriminant analysis, and density estimation. J Am Stat Assoc 97:611–631
16. Xue W, Krasnitz A, Lucito R, Sordella R, VanAelst L, Cordon-Cardo C, Singer S, Kuehnel F, Wigler M, Powers S, Zender L, Lowe SW (2008) DLC1 is a chromosome 8p tumor suppressor whose loss promotes hepatocellular carcinoma. Gene Dev 22:1439–1444

Index

Narendra Wajapeyee (ed.), *Cancer Genomics and Proteomics: Methods and Protocols*, Methods in Molecular Biology, vol. 1176, DOI 10.1007/978-1-4939-0992-6,

MIX
Papier aus verantwortungsvollen Quellen
Paper from responsible sources
FSC® C105338

If you have any concerns about our products,
you can contact us on
ProductSafety@springernature.com

In case Publisher is established outside the EU,
the EU authorized representative is:
Springer Nature Customer Service Center GmbH
Europaplatz 3, 69115 Heidelberg, Germany

Printed by Libri Plureos GmbH
in Hamburg, Germany